Extreme environmental events, such as floods, droughts, rainstorms, and high winds, have severe consequences for human society. How frequently an event of a given magnitude may be expected to occur is of great importance. Planning for weather-related emergencies, design of civil engineering structures, reservoir management, pollution control, and insurance risk calculations, all rely on knowledge of the frequency of these extreme events. Estimation of these frequencies is difficult because extreme events are by definition rare and the data record is often short.

Regional frequency analysis resolves this problem by "trading space for time"; data from several sites are used in estimating event frequencies at any one site. *L*-moments are a recent development within statistics. They form the basis of an elegant mathematical theory in their own right and can be used to facilitate the estimation process in regional frequency analysis. *L*-moment methods are demonstrably superior to those that have been used previously, and are now being adopted by many organizations worldwide.

This book is the first complete account of the *L*-moment approach to regional frequency analysis. It brings together results that previously were scattered among academic journals and also includes much new material. *Regional Frequency Analysis* comprehensively describes the theoretical background to the subject, is rich in practical advice for users, and contains detailed examples that illustrate the approach. This book will be of great value to hydrologists, atmospheric scientists, and civil engineers concerned with environmental extremes.

REGIONAL FREQUENCY ANALYSIS

REGIONAL FREQUENCY ANALYSIS

An Approach Based on *L*-Moments

J. R. M. HOSKING

IBM Research Division
Thomas J. Watson Research Center

and

J. R. WALLIS

School of Forestry and Environmental Studies
Yale University

CAMBRIDGE UNIVERSITY PRESS
Cambridge, New York, Melbourne, Madrid, Cape Town, Singapore,
São Paulo, Delhi, Dubai, Tokyo, Mexico City

Cambridge University Press
The Edinburgh Building, Cambridge CB2 8RU, UK

Published in the United States of America by Cambridge University Press, New York

www.cambridge.org
Information on this title: www.cambridge.org/9780521430456

First published 1997

A catalogue record for this publication is available from the British Library

Library of Congress Cataloguing in Publication Data

Hosking, J. R. M. (Jonathan Richard Morley), 1955–

Regional frequency analysis: an approach based on L- moments/
J. R. M. Hosking and J. R. Wallis

p. cm

Includes bibliographical references (p. -) and index.
ISBN 0-521-43045-3 (hardbound)
1. Distribution (Probability theory) 2. Natural disasters-
-Forecasting–Statistical methods. I. Wallis, James R. II. Title.
QA273.6.H675 1997
519.2'4–DC21 97-4007
CIP

ISBN 978-0-521-43045-6 Hardback
ISBN 978-0-521-01940-8 Paperback

To Evelyn and Lois

Contents

Preface

Many practical problems require the fitting of a probability distribution to a data sample, and in many fields of application the available data consist of not just a single sample but a set of samples drawn from similar probability distributions. It is natural to wonder whether the distribution for one sample can be more accurately estimated by using information not just from that sample but also from the other related samples. In the environmental sciences the data samples are typically measurements of the same kind of data made at different sites, and the process of using data from several sites to estimate the frequency distribution is known as regional frequency analysis. We have developed an approach to regional frequency analysis that is statistically efficient and reasonably straightforward to implement. Our aim in this monograph is to present a complete description of our approach: the specification of all necessary computations, a description of the theoretical statistical background, an assessment of the method's performance in plausible practical situations, recommendations to assist with the subjective decisions that are inevitable in any statistical analysis, and consideration of how to overcome some of the difficulties often encountered in practice. The technical level of exposition is intended to be comprehensible to practitioners with no more than a basic knowledge of probability and statistics, including an understanding of the concepts defined in Sections 2.1–2.3.

The origins of our work can be traced to the early 1970s, when there was a growing awareness among hydrologists that annual maximum streamflow data, although commonly modeled by the Gumbel distribution, often had higher skewness than was consistent with that distribution. Moment statistics were widely used as the basis for identifying and fitting frequency distributions, but to use them effectively required knowledge of their sampling properties in small samples. A massive (for the time) computational effort using simulated data was performed by Wallis, Matalas, and Slack (1974). It revealed some unpleasant properties of moment statistics – high bias and algebraic boundedness. Wallis and others went on to establish the phenomenon of "separation of skewness," which is that for annual maximum streamflow data "the relationship between the mean and the standard deviation of regional estimates of skewness for historical flood sequences is not compatible with the relations derived from several well-known distributions" (Matalas, Slack, and Wallis, 1975). Separation can be explained by "mixed distributions" (Wallis, Matalas, and Slack, 1977) – regional heterogeneity in our present terminology – or if the frequency distribution of streamflow has a longer tail than those of the distributions commonly used in the 1970s. In particular, the Wakeby distribution does not exhibit the phenomenon of separation (Landwehr, Matalas, and Wallis, 1978). The Wakeby distribution was devised by H. A. Thomas Jr. (personal communication to J. R. Wallis, 1976). It is hard to estimate by conventional methods such as maximum likelihood or the method of moments, and the desirability of obtaining closed-form estimates of Wakeby parameters led Greenwood et al. (1979) to devise probability weighted moments. Probability weighted moments were found to perform well for other distributions (Landwehr, Matalas, and Wallis, 1979a; Hosking, Wallis, and Wood, 1985b; Hosking and Wallis, 1987a) but were hard to interpret. Hosking (1990) found that certain linear combinations of probability weighted moments, which he called "*L*-moments," could be interpreted as measures of the location, scale, and shape of probability distributions and formed the basis for a comprehensive theory of the description, identification, and estimation of distributions.

The modern use of the index-flood procedure stems from Wallis (1981, 1982), who used it in conjunction with probability weighted moments and the Wakeby distribution as a method of estimating quantiles in the extreme upper tail of the frequency distribution. Comparative studies showed that this "WAK/PWM" algorithm, and analogs in which other distributions were fitted, outperformed the quantile estimation procedures recommended in the U.K. Flood Studies Report (Hosking, Wallis, and Wood, 1985a) and the U.S. "Bulletin 17" (Wallis and Wood, 1985). Later work investigated the performance of this index-flood procedure in the presence of paleological and historical data (Hosking and Wallis, 1986a,b), regional heterogeneity (Lettenmaier, Wallis, and Wood, 1987), and intersite dependence (Hosking and Wallis, 1988). The practical utility of regional frequency analysis using this index-flood procedure, however, still required subjective judgement at the stages of formation of the regions and choice of an appropriate frequency distribution for each region; statistics to assist with these judgements were developed by Hosking and Wallis (1993).

The foregoing chronology describes our own work, but the complete theory and practice of regional frequency analysis of course involves the work of many other authors, whose contributions we have acknowledged (fairly, we hope!) in the following chapters.

Our work has been greatly facilitated by the unflagging support of the IBM Research Division, in particular that of B. J. Flehinger, W. R. Pulleyblank, and S. Winograd. Credit or blame for the two authors' having met in the first place belongs to J. S. G. McCulloch, then-Director of the Institute of Hydrology in Wallingford, England, and to R. T. Clarke for arranging a visit to the Institute by J. R. Wallis in 1983–84.

It is a pleasure to acknowledge the contributions of several of our colleagues who have read and commented on versions of the manuscript. Our thanks go to D. H. Burn, R. T. Clarke, J. A. Greenwood, N. B. Guttman, C. P. Pearson, and D. W. Reed and to two anonymous reviewers who provided many thought-provoking comments that helped us improve the manuscript. N. B. Guttman provided the data used in Section 9.1 and J. A. Smith sent us the data that he used in Smith (1992). The figures were produced using R. F. Voss's "VossPlot" package (Voss, 1995).

Yorktown Heights, N.Y. J.R.M.H.
August 1996 J.R.W.

Errata

page 64, Table 4.1. In column 4 (No. of sites), row 9, for "2" read "21".

page 74, line 7. For "600 km hr^{-1}" read "1,600 km hr^{-1}".

page 76, line 6. After "range" insert "of".

page 124, line –3. For "$n > 40$" read "$n \geq 100$" (see next item).

page 125, Fig. 7.13. Figure 7.13 shows RMSE of growth curve, not quantiles. A corrected figure is shown below. Observe that, compared with the original version, it is more favorable to regional estimation over at-site estimation.

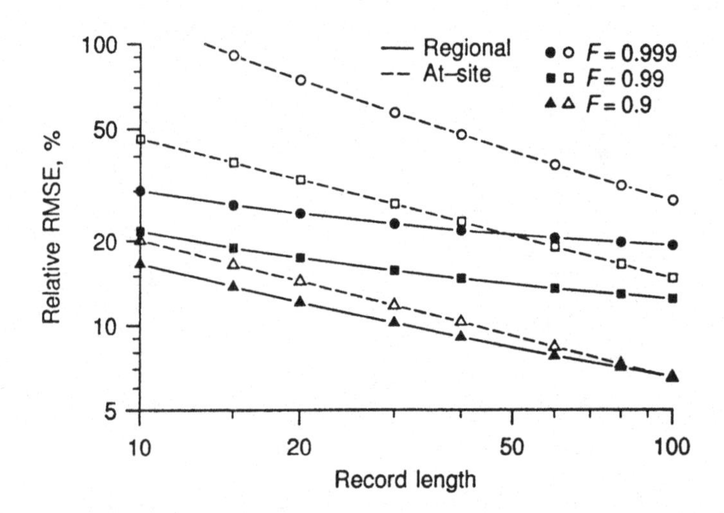

Fig. 7.13, corrected. Regional average relative RMSE of estimated quantiles for heterogeneous regions similar to Region R2 but with different record lengths. Fitted distribution: GEV.

page 203, equation (A.100). The computed value of τ_4 has the wrong sign. The equation should read

$$\tau_4 = (g_1 - 6g_2 + 10g_3 - 5g_4)/(g_1 - g_2)\,.$$

(Note: the values computed by routine `LMRKAP` in the `LMOMENTS` package cited on page 13 are correct.)

pages 220–224. All roman page numbers in the Author Index and the Subject Index are too large by 2. For example, the entry for "separation of skewness" should be xii, not xiv.

1
Regional frequency analysis

1.1 Introduction

Frequency analysis is the estimation of how often a specified event will occur. Estimation of the frequency of extreme events is often of particular importance. Because there are numerous sources of uncertainty about the physical processes that give rise to observed events, a statistical approach to the analysis of data is often desirable. Statistical methods acknowledge the existence of uncertainty and enable its effects to be quantified. Procedures for statistical frequency analysis of a single set of data are well established. It is often the case, however, that many related samples of data are available for analysis. These may, for example, be meteorological or environmental observations of the same variable at different measuring sites, or industrial measurements made on samples of similar products. If event frequencies are similar for the different observed quantities, then more accurate conclusions can be reached by analyzing all of the data samples together than by using only a single sample. In environmental applications this approach is known as *regional frequency analysis*, because the data samples analyzed are typically observations of the same variable at a number of measuring sites within a suitably defined "region." The principles of regional frequency analysis, however, apply whenever multiple samples of similar data are available.

Suppose that observations are made at regular intervals at some site of interest. Let Q be the magnitude of the event that occurs at a given time at a given site. We regard Q as a random quantity (a random variable), potentially taking any value between zero and infinity. The fundamental quantity of statistical frequency analysis is the frequency distribution, which specifies how frequently the possible values of Q occur. Denote by $F(x)$ the probability that the actual value of Q is at most x:

$$F(x) = \Pr[Q \le x]. \tag{1.1}$$

$F(x)$ is the cumulative distribution function of the frequency distribution. Its inverse function $x(F)$, the quantile function of the frequency distribution, expresses the magnitude of an event in terms of its nonexceedance probability F. The quantile of return period T, Q_T, is an event magnitude so extreme that it has probability $1/T$ of being exceeded by any single event. For an extreme high event, in the upper tail of the frequency distribution, Q_T is given by

$$Q_T = x(1 - 1/T) \tag{1.2}$$

or

$$F(Q_T) = 1 - 1/T; \tag{1.3}$$

for an extreme low event, in the lower tail of the frequency distribution, the corresponding relations are $Q_T = x(1/T)$ and $F(Q_T) = 1/T$. The goal of frequency analysis is to obtain a useful estimate of the quantile Q_T for a return period of scientific relevance. This period may be the design life of a structure ($T = 50$ years, say) or some legally mandated design period (e.g., $T = 10000$ years in some dam safety applications). More generally, the goal may be to estimate Q_T for a range of return periods or to estimate the entire quantile function. To be "useful," an estimate should not only be close to the true quantile but should also come with an assessment of how accurate it is likely to be.

If data are available at the site of interest, then the observed data provide a sample of realizations of Q. In many environmental applications the sample size is rarely sufficient to enable quantiles to be reliably estimated. It is generally held that a quantile of return period T can be reliably estimated from a data record of length n only if $T \leq n$. However, in many engineering applications based on annual data (e.g., annual maximum precipitation, streamflow, or windspeed) this condition is rarely satisfied – typically $n < 50$ and $T = 100$ or $T = 1000$. To overcome this problem, several approaches have been devised that use alternative or additional sources of data. This monograph is concerned with one of them – regional frequency analysis.

Regional frequency analysis augments the data from the site of interest by using data from other sites that are judged to have frequency distributions similar to that of the site of interest. If a set of N sites each with n years of record can be found, then one might naively hope that the Nn data values will provide accurate estimates of quantiles as extreme as the Nn-year quantile Q_{Nn}. In practice this is not reasonable; problems arise because frequency distributions at different sites are not exactly identical and because event magnitudes at different sites may not be statistically independent.

Nonetheless, we advocate the use of regional frequency analysis, because we believe that a well-conducted regional analysis will yield quantile estimates accurate enough to be useful in many realistic applications. This conclusion is drawn principally from recent research (Hosking et al., 1985a; Lettenmaier and Potter, 1985; Wallis and Wood, 1985; Lettenmaier et al., 1987; Hosking and Wallis, 1988; Potter and Lettenmaier, 1990) that has investigated the properties of variants of the "regional PWM algorithm," a regional frequency analysis procedure based on statistical quantities called "probability weighted moments" (PWMs) and first used by Greis and Wood (1981) and Wallis (1981, 1982). Cunnane (1988) reviewed twelve different methods of regional frequency analysis and rated the regional PWM algorithm as the best. *L*-moments (Hosking, 1986a, 1990) are statistical quantities that are derived from PWMs and increase the accuracy and ease of use of PWM-based analysis. In this monograph we describe a regional frequency analysis procedure based on *L*-moments, and we show how the procedure can be used to obtain quantile estimates. The next section sets out the principles that underlie our approach.

1.2 Current ideas

Regional frequency analysis has been an established method in hydrology for many years; the index-flood procedure of Dalrymple (1960) is an early example. Several methods recommended by national organizations for general use by hydrologists have a regional component. Bulletin 17 of the U.S. Water Resources Council (1976, 1977, 1981) fits a log-Pearson type III distribution to annual maximum streamflows at a single site, that is, the distribution of log Q is assumed to be Pearson type III. The skewness of the distribution of log Q is estimated by combining a data-based estimate with a value read from a map. The method uses regional information insofar as the mapped values are derived from observed skewness statistics at many sites. It is discussed in more detail in Section 8.3. The method recommended in the U.K. Flood Studies Report (Natural Environment Research Council, 1975) has a strong regional component. It divides the British Isles into eleven regions with region boundaries largely following those of major catchments. The frequency distribution of annual maximum streamflow is assumed to be the same at each gaging site in a region after the streamflow values have been divided by the site mean annual maximum streamflow.

Since these methods were published, research has indicated several ways in which regional frequency analysis can be improved and several principles that are useful for constructing a regional frequency analysis procedure.

Frequency analysis should be robust

Statistical frequency analysis procedures, like virtually all scientific methods, postulate some kind of model for the process that generates the observed data. In most environmental applications the actual data-generating mechanism is so complicated that it is unreasonable to expect the model to be "true," that is, an exact representation of the physical process. A model is at best an approximation. Therefore when fitting a model to the data, any desirable attributes possessed by a model-fitting procedure when the model is true may be irrelevant. Much more important is that the procedure should yield quantile estimates whose accuracy is not seriously degraded when the true physical process deviates from the model's assumptions in a plausible way. A modeling procedure with this property is said to be *robust*.

To assess a frequency analysis procedure, use simulation

To establish the properties of a frequency analysis procedure, or to compare two or more procedures, we recommend the use of Monte Carlo simulation. Though when specifying a model for use in frequency analysis the exact mechanism by which the data are generated may not be known, it can be recognized that some kinds of departure from the model are plausible. For example, the frequency distribution may have a heavier or a lighter tail than the model assumes, and magnitudes of events occurring at the same time at different sites may be correlated. Data can be generated according to whatever pattern of real-world data structure is of concern, and the adequacy of the proposed modeling procedure can be assessed for such data. The advantage of using simulated data for this purpose is that the true quantiles of the frequency distribution are known, so it is easy to judge how well the modeling procedure performs. This is not the case for methods that use only observed data, such as split-sample testing or comparing probability plots of observed samples and fitted distributions.

Regionalization is valuable

Regionalization is the inclusion in frequency analysis of data from sites other than the site at which quantile estimates are required. Because more information is used than in an "at-site" analysis using only a single site's data, there is potential for greater accuracy in the final quantile estimates. But the extra information comes at the price of having to specify the relationships between frequency distributions at different sites. For example, index-flood procedures, such as that described in Section 1.3, assume that frequency distributions at different sites are identical apart from a scale factor, that is, that the sites form a "homogeneous region." Benson (1962) suggested that this assumption was not valid for U.S. streamflow data because the coefficient of variation of the frequency distribution tends to decrease as catchment area increases. Thus there is reason to doubt whether regionalization is worthwhile.

However, research has shown these doubts to be unjustified. Even though a region may be moderately heterogeneous, regional analysis will still yield much more accurate quantile estimates than at-site analysis (Lettenmaier and Potter, 1985; Lettenmaier et al., 1987; Hosking and Wallis, 1988; Potter and Lettenmaier, 1990).

Regions need not be geographical

Regional frequency analysis is advantageous when the sites forming a region have similar frequency distributions. The term "region" suggests a set of neighboring sites, but geographical closeness is not necessarily an indicator of similarity of the frequency distributions. Indeed, for certain kinds of data, some aspects of the frequency distribution can show sharp discontinuities when considered as functions of the location of the site. In the analysis of streamflow data, for example, consider a site downstream of the confluence of two rivers and sites on the two upstream branches. It is plausible that the shape of the frequency distribution could be very different at the three sites. For this reason, maps of regional skewness, as used by Bulletin 17 (U.S. Water Resources Council, 1981), seem likely to be very unreliable.

It is reasonable to identify regions by measuring at each site the variables that are thought to influence the frequency distribution – the "site characteristics" – and then grouping together sites that are adjacent in some suitably defined space of site characteristics. The characteristics used to define this space could be geographical – latitude and longitude, say – but other characteristics may be more directly and physically related to the frequency distribution at the site. In the analysis of streamflow data, for example, such characteristics might include altitude, mean annual precipitation, drainage basin area, soil type, and the location and size of swamps and lakes.

A further advantage of choosing a region that is geographically dispersed rather than compact is that the frequency distributions at the different sites are then less likely to be highly correlated. This reduces the variability of the eventual quantile estimates.

Frequency distributions need not be "textbook" distributions

Lognormal, Pearson type III, and extreme-value type I (Gumbel) are examples of probability distributions for which a fairly thorough mathematical and statistical theory has been developed and which resemble in their general shape what experience suggests a typical frequency distribution should look like for many environmental variables. It is therefore tempting to declare one such "textbook" distribution to be the frequency distribution for fitting to data or to choose a distribution from among a small group of textbook distributions. A problem with this approach is that the sample sizes that are typically available are not so large that the frequency distribution can be unequivocally identified. In particular, failure to detect that the frequency distribution is heavy-tailed, with Q_T increasing rapidly as T increases, will result in severe underestimation of extreme quantiles. Several authors have found evidence that frequency distributions of environmental data can be heavy-tailed (Houghton, 1978; Landwehr et al., 1978; Rossi, Fiorentino, and Versace, 1984; Ahmad, Sinclair, and Werritty, 1988). It is therefore wise to consider as candidate frequency distributions a wide range of moderate- and heavy-tailed distributions or to use a distribution with enough free parameters that it can mimic a wide range of plausible frequency distributions. The Wakeby distribution (Houghton, 1978), with five parameters, is one such "mimic-everything" distribution.

L-moments are useful summary statistics

Most regional frequency analysis procedures attempt to fit to the data a distribution whose form is specified apart from a finite number of undetermined parameters. Sample moment statistics, particularly skewness and kurtosis, are often used to judge the closeness of an observed sample to a postulated distribution. However, these statistics are unsatisfactory. They are algebraically bounded, with bounds dependent on sample size, and in many small or moderate samples it is unusual for sample skewness and kurtosis to take values anywhere near the population values.

We recommend an alternative approach based on the "*L*-moment" statistics described in Chapter 2. These are analogous to the conventional moments but can be estimated by linear combinations of the elements of an ordered sample, that is, by *L*-statistics. *L*-moments have the theoretical advantages over conventional moments of being able to characterize a wider range of distributions and, when estimated from a sample, of being more robust to the presence of outliers in the data. Experience also shows that, compared with conventional moments, *L*-moments are less subject to bias in estimation. Examples and further comparisons of moments and *L*-moments are given in Section 2.9.

1.3 An index-flood procedure

Index-flood procedures are a convenient way of pooling summary statistics from different data samples. The term "index flood" arose because early applications of the procedure were to flood data in hydrology (e.g., Dalrymple, 1960), but the method can be used with any kind of data.

Suppose that data are available at N sites, with site i having sample size n_i and observed data Q_{ij}, $j = 1, \dots, n_i$. Let $Q_i(F), 0 < F < 1$, be the quantile function of the frequency distribution at site i. The key assumption of an index-flood procedure is that the sites form a *homogeneous region*, that is, that the frequency distributions of the N sites are identical apart from a site-specific scaling factor, the *index flood*.

We may then write

$$Q_i(F) = \mu_i q(F), \qquad i = 1, \ldots, N. \tag{1.4}$$

Here μ_i is the index flood. We shall take it to be the mean of the at-site frequency distribution, though any location parameter of the distribution may be used instead. Smith (1989), for example, used the 90% quantile $Q_i(0.9)$. The remaining factor in (1.4), $q(F)$, is the *regional growth curve*, a dimensionless quantile function common to every site. It is the quantile function of the *regional frequency distribution*, the common distribution of the Q_{ij}/μ_i.

The index flood is naturally estimated by $\hat{\mu}_i = \bar{Q}_i$, the sample mean of the data at site i. Other location estimators such as the median or a trimmed mean could be used instead.

The dimensionless rescaled data $q_{ij} = Q_{ij}/\hat{\mu}_i$, $j = 1, \ldots, n_i$, $i = 1, \ldots, N$, are the basis for estimating the regional growth curve $q(F)$, $0 < F < 1$. It is usually assumed that the form of $q(F)$ is known apart from p undetermined parameters $\theta_1, \ldots, \theta_p$, so we write $q(F)$ as $q(F; \theta_1, \ldots, \theta_p)$. For example, these parameters may be the coefficient of variation and the skewness of the distribution, or the *L*-moment ratios τ, $\tau_3, \ldots,$ defined in Section 2.4. The mean of the regional frequency distribution is not an unknown parameter, because by taking μ_i in (1.4) to be the mean of the frequency distribution at site i we ensure that the regional frequency distribution has mean 1. In our approach the parameters are estimated separately at each site, the site-i estimate of θ_k being denoted by $\hat{\theta}_k^{(i)}$. These at-site estimates are combined to give regional estimates:

$$\hat{\theta}_k^{\mathrm{R}} = \sum_{i=1}^{N} n_i \hat{\theta}_k^{(i)} \Big/ \sum_{i=1}^{N} n_i \,. \tag{1.5}$$

This is a weighted average, with the site-i estimate given weight proportional to n_i because for regular statistical models the variance of $\hat{\theta}_k^{(i)}$ is inversely proportional to n_i. Substituting these estimates into $q(F)$ gives the estimated regional growth curve $\hat{q}(F) = q(F; \hat{\theta}_1^{\mathrm{R}}, \ldots, \hat{\theta}_p^{\mathrm{R}})$. This method of obtaining regional estimates is essentially that of Wallis (1981), except that the weighting proportional to n_i is a later addition, suggested by Wallis (1982). Somewhat different methods were used by Dalrymple (1960) and the Natural Environment Research Council (1975).

The quantile estimates at site i are obtained by combining the estimates of μ_i and $q(F)$:

$$\hat{Q}_i(F) = \hat{\mu}_i \hat{q}(F). \tag{1.6}$$

This index-flood procedure makes the following assumptions.

(i) Observations at any given site are identically distributed.
(ii) Observations at any given site are serially independent.
(iii) Observations at different sites are independent.
(iv) Frequency distributions at different sites are identical apart from a scale factor.
(v) The mathematical form of the regional growth curve is correctly specified.

The first two assumptions are plausible for many kinds of data, particularly for annual totals or extremes, which are free from seasonal variations. It is a basic assumption of most methods of frequency analysis that the events observed in the past are likely to be typical of what may be expected in the future. This assumption may be undermined when obvious sources of time trends are present; frequency distributions for streamflow data, for example, are affected by changes in land use and by artificial regulation of the flow. When sites affected by such obvious sources of nonstationarity are removed from the data set, the assumption of identical distributions for a site's observations is often reasonable.

The effect of serial dependence on at-site frequency analysis has been investigated by Landwehr et al. (1979a) and McMahon and Srikanthan (1982). They considered frequency distributions of extreme-value type I and log-Pearson type III, respectively, and found that serial dependence caused a small amount of bias and a small increase in the standard error of quantile estimates. We conclude that a small amount of serial dependence in annual data series has little effect on the quality of quantile estimates. If trends, periodic variation, or serial dependence are present to a large extent in the data, some kind of time-series analysis is likely to be more appropriate than the time-independent frequency analysis considered here.

The last three assumptions are unlikely to be satisfied by environmental data. Correlation between nearby sites may be expected for many kinds of data. Meteorological events such as storms and droughts typically affect an area large enough to contain more than one measuring site, and the event magnitudes at neighboring sites are therefore likely to be positively correlated. The last two assumptions will never be exactly valid in practice. At best they may be approximately attained, by careful selection of the sites that are to be regarded as forming a region and by careful choice of a frequency distribution that is consistent with the data. Therefore an index-flood procedure can be appropriate only if it is robust to physically plausible departures from these three assumptions. Recent research (Hosking et al., 1985a; Lettenmaier and Potter, 1985; Wallis and Wood, 1985; Lettenmaier et al., 1987; Hosking and Wallis, 1988) has shown that it is possible to construct index-flood procedures that yield suitably robust and accurate quantile estimates.

The definition of a homogeneous region and the relation (1.4) between the at-site quantile functions are appropriate when the quantity of interest, Q, can take only positive values. Some quantities, such as temperature measured on the Celsius scale, can take both positive and negative values. The index-flood procedure described here is not appropriate for such data, but analogous methods could be developed based on suitable modifications of Eq. (1.4). For example, if instead of Eq. (1.4) the frequency distributions in a region could be described by the location-scale model $Q_i(F) = \mu_i + \sigma_i q(F)$, the regional shape estimation procedure described in Section 8.2 would be appropriate.

An alternative approach to regional estimation is to model $\log Q$ rather than Q, basing the analysis on logarithmically transformed data. Taking logarithms in Eq. (1.4) gives

$$\log Q_i(F) = \log \mu_i + \log q(F) . \tag{1.7}$$

The index flood enters as an additive term, which makes some aspects of the analysis easier. For example, if unbiased estimators of $\log \mu_i$ and $\log q(F)$ can be found, their sum will be an unbiased estimator of $\log Q_i(F)$. The disadvantage of using log-transformed data is that low data values may become low outliers after logarithmic transformation and have an undue influence on the estimates. In applications in which estimation of quantiles in the upper tail of the distribution is of principal importance, it is particularly unfortunate for low data values to have a strong effect on the upper tail of the estimated frequency distribution. For this reason we generally prefer to work with the original untransformed data.

1.4 Steps in regional frequency analysis

Given that data are available at a large number of sites and that quantile estimates are required at each site, regional frequency analysis using an index-flood procedure will involve the four steps outlined below.

Step 1. Screening of the data

As with any statistical analysis, the first stage of regional frequency analysis is a close inspection of the data. Gross errors and inconsistencies should be eliminated and a check made that the data are homogeneous (stationary) over time. External information can be useful here, especially information about methods of data collection and measurement and about any changes over time that may have affected the frequency distribution at any site.

Step 2. Identification of homogeneous regions

The next step in regional frequency analysis is the assignment of the sites to regions. A *region*, a set of sites whose frequency distributions are (after appropriate scaling) considered to be approximately the same, is the fundamental unit of regional frequency analysis. We do not suppose that the sites can be divided into regions within which the homogeneity criterion (1.4) is exactly satisfied. Approximate homogeneity is sufficient to ensure that regional frequency analysis is much more accurate than at-site analysis.

As noted in Section 1.2, regions need not be geographical, but should instead consist of sites having similar values of those site characteristics that determine the frequency distribution. Suitable site characteristics depend on the kind of data being analyzed. Latitude and longitude are also site characteristics and may be used as surrogates for unmeasured characteristics that vary smoothly with location. The homogeneity of a proposed region should be tested by calculating summary statistics of the at-site data and comparing the between-site variability of these statistics with what would be expected of a homogeneous region. L-moments are suitable statistics for this purpose.

Step 3. Choice of a frequency distribution

After a region has been identified, the final stage in the specification of the statistical model is the choice of an appropriate regional frequency distribution, $q(F)$ in Eq. (1.4). This is a common statistical problem, often solved by applying a goodness-of-fit test, a procedure that involves computing summary statistics from the data and testing whether their values are consistent with what would be expected if the data were a random sample from some postulated distribution. This approach can be used in regional frequency analysis, but two extra considerations apply. First, the available data are not a single random sample but a set of samples from the different sites; and second, the chosen distribution should not merely fit the data well but should also yield quantile estimates that are robust to physically plausible deviations of the true frequency distribution from the chosen frequency distribution.

Step 4. Estimation of the frequency distribution

Estimation of the regional frequency distribution can be achieved by estimating the distribution separately at each site and combining the at-site estimates to give a regional average, as described in Section 1.3. An efficient method of doing this is to combine the at-site L-moment statistics via the weighted average (1.5); we call this method the *regional L-moment algorithm.*

There are two important situations in which the foregoing procedure must be modified or extended.

First, there may be one site of special interest, such as a nuclear power plant or an actual or proposed dam site, where the aim of the analysis is to obtain quantile estimates for this site. In this case special care should be taken to make the site typical of the region to which it is assigned. So far as is possible, the site characteristics of the site of interest should be typical of those of the other sites in its region and should not be at either extreme of the range of values of the site characteristics. This is to reduce the bias in quantile estimates that can occur at sites that are not typical of the region as a whole.

Second, quantile estimates may be required at one or more *ungaged sites*, where no data have been observed. On the basis of its site characteristics, an ungaged site can be assigned to one of the regions identified for the gaged sites. This gives an estimate of the regional growth curve at the ungaged site. There remains only the problem of estimating the index flood, usually the mean μ of the at-site frequency distribution, at ungaged sites. The most reasonable approach is to regard μ as being a function of site characteristics and to calibrate the relationship between μ and site characteristics by using data from the gaged sites. This is discussed in Section 8.4.

1.5 Outline of the monograph

The following chapters expand on the approach described in the foregoing sections.

Chapter 2 contains a general introduction to *L*-moments, which form the basis of our statistical methods.

Chapters 3 through 6 describe the four steps in regional frequency analysis listed in Section 1.4. The first three steps involve subjective judgement; *L*-moments are used to construct statistics that provide objective backing for these judgements.

Chapter 3 is concerned with the initial screening of data. *L*-moments can be used to construct a discordancy measure, described in Section 3.2. This identifies unusual sites, those whose at-site sample *L*-moment ratios are markedly different from those of the other sites in the data set. The discordancy measure provides an initial screening of the data and indicates sites where the data may merit close examination.

Chapter 4 is concerned with the construction and testing of homogeneous regions. Section 4.1 surveys methods of forming regions. After an initial set of regions has been obtained, there is a need to test whether a proposed region is acceptably close to homogeneous. This can be done by calculating summary statistics of the at-site data and comparing the between-site variability of these statistics with what would be expected of a homogeneous region. Section 4.3 describes a heterogeneity measure that performs this test. The summary statistics that it uses are the sample *L*-moments.

Chapter 5 is concerned with the choice of an appropriate frequency distribution. Section 5.2 describes a goodness-of-fit measure. This tests whether a candidate distribution gives a good fit to a region's data: specifically, whether there is a statistically significant difference between the regional average *L*-moments and those of the fitted distribution.

Table 1.1. *Conversions between imperial and metric units.*

Imperial unit	Metric equivalent
inch	1 in = 25.4 mm
foot	1 ft = 0.3048 m
mile	1 mi = 1.609 km
square mile	1 mi^2 = 2.590 km^2
cubic foot	1 ft^3 = 0.02832 m^3

Chapter 6 is concerned with the estimation of regional and at-site quantiles. Section 6.2 describes the regional *L*-moment algorithm. Section 6.4 describes a method of assessing the accuracy of estimated quantiles.

Chapter 7 studies the performance of the regional *L*-moment algorithm for a wide range of possible regions.

Chapter 8 discusses some topics peripheral to our main concerns, including variants of the basic index-flood procedure for regional frequency analysis, quantile estimation at ungaged sites and the use of historical information.

Chapter 9 presents two detailed examples of regional frequency analysis, illustrating all the steps listed in Section 1.4.

The appendix contains specifications of the cumulative distribution functions and quantile functions, and the relations between parameters and *L*-moments, for a selection of distributions useful in regional frequency analysis.

The data used in the examples are of U.S. origin and are measured in imperial units. Conversions to metric units are given in Table 1.1.

The methods described in this monograph involve a considerable amount of computation but are comfortably within the scope of current personal computers and mainframes. The numerical methods have been programmed as Fortran routines. Hosking (1996) documents routines that perform the following calculations:

- computation of the cumulative distribution function and quantile function of the distribution, the *L*-moments given the parameters, and the parameters given the low-order *L*-moments for each of the distributions discussed in the appendix – except the uniform distribution;
- computation of the probability weighted moments and *L*-moments of a data sample;
- cluster analysis using Ward's method;
- computation of the discordancy, heterogeneity, and goodness-of-fit measures described in Chapters 3–5;

- fitting a distribution to a regional data set using the regional L-moment algorithm described in Chapter 6; and
- assessment of the accuracy of regional estimates using the procedure described in Section 6.4.

The LMOMENTS package contains Fortran-77 source code for these routines. It is freely available electronically from the authors, by request from hosking@watson.ibm.com, or from the Statlib software repository, http://lib.stat.cmu.edu/general/lmoments.

2
L-moments

2.1 Probability distributions

Let X be a random variable, taking values that are real numbers. The relative frequency with which these values occur defines the *frequency distribution* or probability distribution of X and is specified by the *cumulative distribution function*

$$F(x) = \Pr[X \le x], \tag{2.1}$$

where $\Pr[\mathcal{A}]$ denotes the probability of the event $\mathcal{A}$. $F(x)$ is an increasing function of x, and $0 \le F(x) \le 1$ for all x. We shall normally be concerned with continuous random variables, for which $\Pr[X = t] = 0$ for all t, that is, no single value has nonzero probability. In this case, $F(.)$ is a continuous function and has an inverse function $x(.)$, the *quantile function* of X. Given any u, $0 < u < 1$, $x(u)$ is the unique value that satisfies

$$F(x(u)) = u. \tag{2.2}$$

For any probability p, $x(p)$ is the *quantile* of nonexceedance probability p, that is, the value such that the probability that X does not exceed $x(p)$ is p. The goal of frequency analysis is accurate estimation of the quantiles of the distribution of some random variable. In engineering and environmental applications a quantile is often expressed in terms of its *return period*, defined in Section 1.1.

If $F(x)$ is differentiable, its derivative $f(x) = \frac{d}{dx}F(x)$ is the *probability density function* of X.

The *expectation* of the random variable X is defined to be

$$\mathrm{E}(X) = \int_{-\infty}^{\infty} x\, dF(x) = \int_{-\infty}^{\infty} x\, f(x)\, dx \tag{2.3}$$

provided that this integral exists. We may also write, via the transformation $u = F(x)$,

$$\mathrm{E}(X) = \int_0^1 x(u)\,du\,. \tag{2.4}$$

A function $g(X)$ of a random variable is itself a random variable and has expectation

$$\mathrm{E}\{g(X)\} = \int_{-\infty}^{\infty} g(x)\,dF(x) = \int_{-\infty}^{\infty} g(x)\,f(x)\,dx = \int_0^1 g(x(u))\,du\,. \tag{2.5}$$

The dispersion of the values taken by the random variable X can be measured by the *variance* of X,

$$\mathrm{var}(X) = \mathrm{E}[\{X - \mathrm{E}(X)\}^2]\,. \tag{2.6}$$

We shall occasionally use measures of the tendency of two random variables X and Y to take large values simultaneously. This can be measured by the *covariance* of X and Y,

$$\mathrm{cov}(X, Y) = \mathrm{E}[\{X - \mathrm{E}(X)\}\{Y - \mathrm{E}(Y)\}]\,. \tag{2.7}$$

The *correlation* between X and Y,

$$\mathrm{corr}(X, Y) = \mathrm{cov}(X, Y)/\{\mathrm{var}(X)\,\mathrm{var}(Y)\}^{1/2}\,, \tag{2.8}$$

is a dimensionless analog of covariance, taking values between -1 and $+1$.

2.2 Estimators

In practice it is often assumed that the distribution of some physical quantity is exactly known apart from a finite set of parameters $\theta_1, \ldots, \theta_p$. When needed for clarity, we write the quantile function of a distribution with p unknown parameters as $x(u; \theta_1, \ldots, \theta_p)$. In most applications the unknown parameters include a location parameter and a scale parameter. A parameter ξ of a distribution is a *location parameter* if the quantile function of the distribution satisfies

$$x(u; \xi, \theta_2, \ldots, \theta_p) = \xi + x(u; 0, \theta_2, \ldots, \theta_p)\,. \tag{2.9}$$

A parameter α of a distribution is a *scale parameter* if the quantile function of the distribution satisfies

$$x(u; \alpha, \theta_2, \ldots, \theta_p) = \alpha \times x(u; 1, \theta_2, \ldots, \theta_p), \tag{2.10}$$

or, if the distribution also has a location parameter ξ,

$$x(u; \xi, \alpha, \theta_3, \ldots, \theta_p) = \xi + \alpha \times x(u; 0, 1, \theta_3, \ldots, \theta_p). \tag{2.11}$$

The unknown parameters are estimated from the observed data. Given a set of data, a function $\hat{\theta}$ of the data values may be chosen as an estimator of θ. The estimator $\hat{\theta}$ is a random variable and has a probability distribution. The goodness of $\hat{\theta}$ as an estimator of θ depends on how close $\hat{\theta}$ typically is to θ. The deviation of $\hat{\theta}$ from θ may be decomposed into bias – a tendency to give estimates that are consistently higher or lower than the true value – and variability – the random deviation of the estimate from the true value that occurs even for estimators that have no bias.

Common measures of the performance of an estimator $\hat{\theta}$ are its *bias* and *root mean square error* (RMSE), defined by

$$\text{bias}(\hat{\theta}) = \mathrm{E}(\hat{\theta} - \theta), \qquad \text{RMSE}(\hat{\theta}) = \{\mathrm{E}(\hat{\theta} - \theta)^2\}^{1/2}. \tag{2.12}$$

We say that $\hat{\theta}$ is *unbiased* if $\text{bias}(\hat{\theta}) = 0$, that is, if $\mathrm{E}(\hat{\theta}) = \theta$. Different unbiased estimators of the same parameter may be compared in terms of their variance; the ratio $\text{var}(\hat{\theta}^{(1)}) / \text{var}(\hat{\theta}^{(2)})$ is the *efficiency* of the estimator $\hat{\theta}^{(2)}$ relative to the estimator $\hat{\theta}^{(1)}$. We have

$$\text{RMSE}(\hat{\theta}) = [\{\text{bias}(\hat{\theta})\}^2 + \text{var}(\hat{\theta})]^{1/2}, \tag{2.13}$$

showing that RMSE combines the bias and variability of $\hat{\theta}$ to give an overall measure of estimation accuracy. In regular statistical problems involving estimation based on a sample of size n, both the bias and variance of $\hat{\theta}$ are asymptotically proportional to n^{-1} for large n (see, e.g., Cox and Hinkley, 1974, Section 9.2). The RMSE of $\hat{\theta}$ is therefore typically proportional to $n^{-1/2}$.

Both the bias and RMSE of $\hat{\theta}$ have the same units of measurement as the parameter θ. It is convenient to express bias and RMSE as ratios with respect to the parameter itself. We thereby obtain dimensionless measures, the *relative bias*, $\text{bias}(\hat{\theta})/\theta$, and the *relative RMSE*, $\text{RMSE}(\hat{\theta})/\theta$. These are the quantities that we primarily use in Chapter 7 to compare different estimators in regional frequency analysis.

2.3 Moments

The shape of a probability distribution has traditionally been described by the moments of the distribution. The moments are the *mean*

$$\mu = \mathrm{E}(X) \tag{2.14}$$

and the higher moments

$$\mu_r = \mathrm{E}(X - \mu)^r, \qquad r = 2, 3, \ldots. \tag{2.15}$$

The mean is the center of location of the distribution. The dispersion of the distribution about its center is measured by the *standard deviation*,

$$\sigma = \mu_2^{1/2} = \{\mathrm{E}(X - \mu)^2\}^{1/2}, \tag{2.16}$$

or the variance, $\sigma^2 = \mathrm{var}(X)$. The *coefficient of variation* (CV), $C_v = \sigma/\mu$, expresses the dispersion of a distribution as a proportion of the mean. Dimensionless higher moments $\mu_r/\mu_2^{r/2}$ are also used, particularly the *skewness*

$$\gamma = \mu_3/\mu_2^{3/2} \tag{2.17}$$

and the *kurtosis*

$$\kappa = \mu_4/\mu_2^2. \tag{2.18}$$

Analogous quantities can be computed from a data sample $x_1, x_2, \ldots, x_n$. The *sample mean*

$$\bar{x} = n^{-1} \sum_{i=1}^{n} x_i \tag{2.19}$$

is the natural estimator of μ. The higher sample moments

$$m_r = n^{-1} \sum_{i=1}^{n} (x_i - \bar{x})^r \tag{2.20}$$

are reasonable estimators of the μ_r, but are not unbiased. Unbiased estimators are often used. In particular, σ^2, μ_3 and the fourth cumulant $\kappa_4 = \mu_4 - 3\mu_2^2$ are unbiasedly estimated by

$$s^2 = (n-1)^{-1} \sum_{i=1}^{n} (x_i - \bar{x})^2 , \tag{2.21}$$

$$\tilde{m}_3 = \frac{n^2}{(n-1)(n-2)} m_3 , \tag{2.22}$$

$$\tilde{k}_4 = \frac{n^2}{(n-2)(n-3)} \left\{ \left(\frac{n+1}{n-1} \right) m_4 - 3m_2^2 \right\} , \tag{2.23}$$

respectively. The sample standard deviation, $s = \sqrt{s^2}$, is an estimator of σ but is not unbiased. The sample estimators of CV, skewness and kurtosis are, respectively,

$$\hat{C}_v = s/\bar{x}, \qquad g = \tilde{m}_3/s^3, \qquad k = \tilde{k}_4/s^4 + 3. \tag{2.24}$$

Moment estimators have some undesirable properties. The estimators g and k can be severely biased, as noted by many authors and investigated in detail by Wallis et al. (1974). Indeed g and k have algebraic bounds that depend on the sample size; for a sample of size n the bounds are

$$|g| \le n^{1/2} \qquad \text{and} \qquad k \le n+3. \tag{2.25}$$

These results follow from bounds on $m_3/m_2^{3/2}$ and m_4/m_2^2 given by Wilkins (1944) – whose work has apparently been overlooked by other authors who have subsequently rediscovered similar bounds – and Dalén (1987), respectively. Thus if a distribution is sufficiently skew, it may be impossible for this skewness to be reflected in a sample of fixed size. For example, a two-parameter lognormal distribution, with cumulative distribution function given by Eq. (A.67) with $\sigma = 1$, has skewness 6.91, but a sample of size 20 drawn from this distribution cannot have sample skewness larger than 4.47, or 65% of the population value.

Inferences based on sample moments of skew distributions are therefore likely to be very unreliable. A more satisfactory set of measures of distributional shape is obtained from *L*-moments, described in the next section.

2.4 *L*-moments of probability distributions

L-moments are an alternative system of describing the shapes of probability distributions. Historically they arose as modifications of the "probability weighted moments" of Greenwood et al. (1979).

Probability weighted moments of a random variable X with cumulative distribution function $F(\,.\,)$ were defined by Greenwood et al. (1979) to be the quantities

$$M_{p,r,s} = \mathrm{E}\left[X^p\{F(X)\}^r\{1 - F(X)\}^s\right]. \tag{2.26}$$

Particularly useful special cases are the probability weighted moments $\alpha_r = M_{1,0,r}$ and $\beta_r = M_{1,r,0}$. For a distribution that has a quantile function $x(u)$, Eqs. (2.5) and (2.26) give

$$\alpha_r = \int_0^1 x(u)(1-u)^r\,du\,, \qquad \beta_r = \int_0^1 x(u)u^r\,du\,. \tag{2.27}$$

These equations may be contrasted with the definition of the ordinary moments, which may be written as

$$\mathrm{E}(X^r) = \int_0^1 \{x(u)\}^r\,du\,. \tag{2.28}$$

Conventional moments involve successively higher powers of the quantile function $x(u)$, whereas probability weighted moments involve successively higher powers of u or $1 - u$ and may be regarded as integrals of $x(u)$ weighted by the polynomials u^r or $(1-u)^r$.

The probability weighted moments α_r and β_r have been used as the basis of methods for estimating parameters of probability distributions by Landwehr, Matalas and Wallis (1979a,b), Greis and Wood (1981), Wallis (1981, 1982), Hosking, Wallis and Wood (1985b) and Hosking and Wallis (1987a). However, they are difficult to interpret directly as measures of the scale and shape of a probability distribution. This information is carried in certain linear combinations of the probability weighted moments. For example, estimates of scale parameters of distributions are multiples of $\alpha_0 - 2\alpha_1$ or $2\beta_1 - \beta_0$ (Landwehr, Matalas and Wallis 1979a; Hosking, Wallis and Wood, 1985b). The skewness of a distribution can be measured by $6\beta_2 - 6\beta_1 + \beta_0$ (Stedinger, 1983). These linear combinations arise naturally from integrals of $x(u)$ weighted not by the polynomials u^r or $(1-u)^r$ but by a set of orthogonal polynomials.

We define polynomials $P_r^*(u)$, $r = 0, 1, 2, \ldots$, as follows

(i) $P_r^*(u)$ is a polynomial of degree r in u.
(ii) $P_r^*(1) = 1$.
(iii) $\int_0^1 P_r^*(u)P_s^*(u)\,du = 0$ if $r \neq s$.

Condition (iii) is the orthogonality condition. These conditions define the *shifted Legendre polynomials* ("shifted," because the ordinary Legendre polynomials $P_r(u)$

are defined to be orthogonal on the interval $-1 \le u \le +1$, not $0 \le u \le 1$). The polynomials have the explicit form

$$P_r^*(u) = \sum_{k=0}^{r} p_{r,k}^* u^k, \tag{2.29}$$

where

$$p_{r,k}^* = (-1)^{r-k} \binom{r}{k} \binom{r+k}{k} = \frac{(-1)^{r-k}(r+k)!}{(k!)^2 \, (r-k)!}. \tag{2.30}$$

For a random variable X with quantile function $x(u)$, we now define the L-moments of X to be the quantities

$$\lambda_r = \int_0^1 x(u) \, P_{r-1}^*(u) \, du \, . \tag{2.31}$$

In terms of probability weighted moments, L-moments are given by

$$\lambda_1 = \alpha_0 \qquad = \beta_0 \, , \tag{2.32}$$

$$\lambda_2 = \alpha_0 - 2\alpha_1 \qquad = 2\beta_1 - \beta_0 \, , \tag{2.33}$$

$$\lambda_3 = \alpha_0 - 6\alpha_1 + 6\alpha_2 \qquad = 6\beta_2 - 6\beta_1 + \beta_0 \, , \tag{2.34}$$

$$\lambda_4 = \alpha_0 - 12\alpha_1 + 30\alpha_2 - 20\alpha_3 = 20\beta_3 - 30\beta_2 + 12\beta_1 - \beta_0 \, , \tag{2.35}$$

and in general

$$\lambda_{r+1} = (-1)^r \sum_{k=0}^{r} p_{r,k}^* \alpha_k = \sum_{k=0}^{r} p_{r,k}^* \beta_k \, . \tag{2.36}$$

It is convenient to define dimensionless versions of L-moments; this is achieved by dividing the higher-order L-moments by the scale measure λ_2. We define the *L-moment ratios*

$$\tau_r = \lambda_r / \lambda_2, \qquad r = 3, 4, \ldots . \tag{2.37}$$

L-moment ratios measure the shape of a distribution independently of its scale of measurement.

We also define the L-CV

$$\tau = \lambda_2/\lambda_1. \tag{2.38}$$

This quantity is analogous to the ordinary coefficient of variation, C_v. (L-CV is not an abbreviation of "L-coefficient of variation": in words it would be more properly described as a "coefficient of L-variation.")

2.5 *L*-moments and order statistics

An intuitive justification for L-moments can be obtained by considering linear combinations of the observations in a sample of data that has been arranged in ascending order. Consider the measurement of the shape of a distribution, given a small sample drawn from the distribution. Denote by $X_{k:n}$ the kth smallest observation from a sample of size n, so that the ordered sample is $X_{1:n} \le X_{2:n} \le \cdots \le X_{n:n}$.

A sample of size 1 is the single observation $X_{1:1}$. It contains information about the location of the distribution. If the distribution is shifted towards larger values, then we would expect to observe larger values of $X_{1:1}$. See Figure 2.1.

A sample of size 2 contains two observations, $X_{1:2}$ and $X_{2:2}$. The sample contains information about the scale, or dispersion, of the distribution. If the distribution is tightly bunched around a central value, then the two observations will tend to be close together. If the distribution is widely dispersed, then the two observations will typically be far apart. See Figure 2.2. Thus the difference between the two observations, $X_{2:2} - X_{1:2}$, is a measure of the scale of the distribution.

A sample of size 3, $X_{1:3} \le X_{2:3} \le X_{3:3}$, also contains information about the skewness of the distribution. If the distribution is symmetric about a central value, then the two extreme observations will typically be approximately equidistant from the central observation; that is, we will have $X_{3:3} - X_{2:3} \approx X_{2:3} - X_{1:3}$ or $X_{3:3} - 2X_{2:3} + X_{1:3} \approx 0$. If the distribution is skewed to the right, so that the upper tail is heavier than the lower tail, then typically $X_{3:3} - X_{2:3}$ will be larger than $X_{2:3} - X_{1:3}$, and so $X_{3:3} - 2X_{2:3} + X_{1:3}$ will be positive. See Figure 2.3. Similarly, if the distribution is skewed to the left, $X_{3:3} - 2X_{2:3} + X_{1:3}$ will typically be negative. Thus $X_{3:3} - 2X_{2:3} + X_{1:3}$, the central second difference of the ordered sample, is a measure of the skewness of the distribution.

For a sample of size 4 we are similarly led to consider the central third difference of the ordered sample, $X_{4:4} - 3X_{3:4} + 3X_{2:4} - X_{1:4}$. Writing this as $(X_{4:4} - X_{1:4}) - 3(X_{3:4} - X_{2:4})$, we see that it measures how much further apart the two extreme values of the sample are than the two central values. If the distribution has a flat density function, then the sample values will typically be approximately equally spaced and the central third difference will be close to zero. If the distribution has a high central peak and long tails, then the central third difference is typically large. See Figure 2.4. Thus $X_{4:4} - 3X_{3:4} + 3X_{2:4} - X_{1:4}$ is a measure of the kurtosis of the distribution.

We have seen that certain linear combinations of the elements of an ordered sample contain information about the location, scale, and shape of the distribution from which the sample was drawn. *L*-moments are defined to be the expected values of these linear combinations, multiplied for numerical convenience by scalar constants. The "*L*" in *L*-moments emphasizes the construction of *L*-moments from linear combinations of order statistics. The *L*-moments of a probability distribution are defined by

$$\lambda_1 = \mathrm{E}(X_{1:1}), \tag{2.39}$$

$$\lambda_2 = \tfrac{1}{2}\,\mathrm{E}(X_{2:2} - X_{1:2}), \tag{2.40}$$

$$\lambda_3 = \tfrac{1}{3}\,\mathrm{E}(X_{3:3} - 2X_{2:3} + X_{1:3}), \tag{2.41}$$

$$\lambda_4 = \tfrac{1}{4}\,\mathrm{E}(X_{4:4} - 3X_{3:4} + 3X_{2:4} - X_{1:4}), \tag{2.42}$$

and in general

$$\lambda_r = r^{-1} \sum_{j=0}^{r-1} (-1)^j \binom{r-1}{j} \mathrm{E}(X_{r-j:r}). \tag{2.43}$$

The two definitions (2.31) and (2.43) are consistent. The expectation of an order statistic can be written

$$\mathrm{E}(X_{r:n}) = \frac{n!}{(r-1)!\,(n-r)!} \int_0^1 x(u)\, u^{r-1} (1-u)^{n-r}\, du\,, \tag{2.44}$$

whence in Eq. (2.43) λ_r can be written as an integral of $x(u)$ multiplied by a polynomial in u; this polynomial can be shown to be $P^*_{r-1}(u)$ (Hosking, 1990, p. 106).

2.6 Properties of *L*-moments

The *L*-moments λ_1 and λ_2, the *L*-CV τ and the *L*-moment ratios τ_3 and τ_4 are the most useful quantities for summarizing probability distributions. Their most important properties are as follows (proofs are given in Hosking (1989, 1990)):

Existence. If the mean of the distribution exists, then all of the *L*-moments exist.

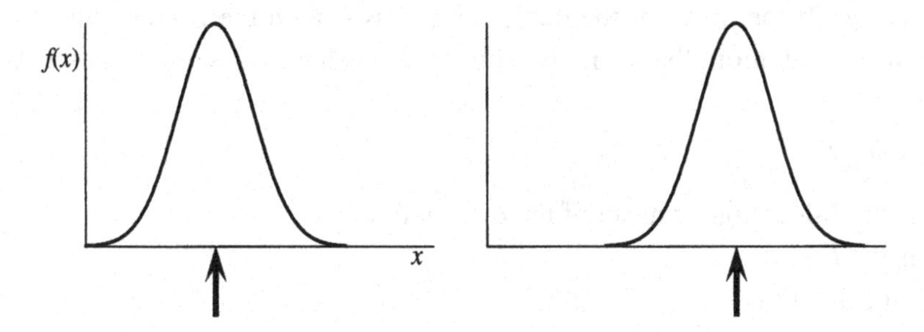

Fig. 2.1. Definition sketch for first L-moment.

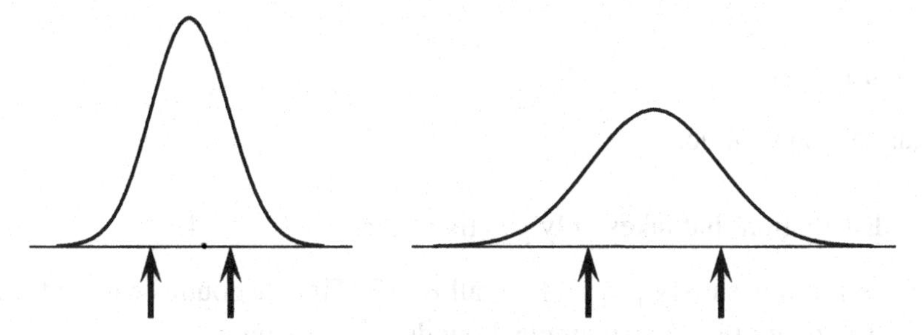

Fig. 2.2. Definition sketch for second L-moment.

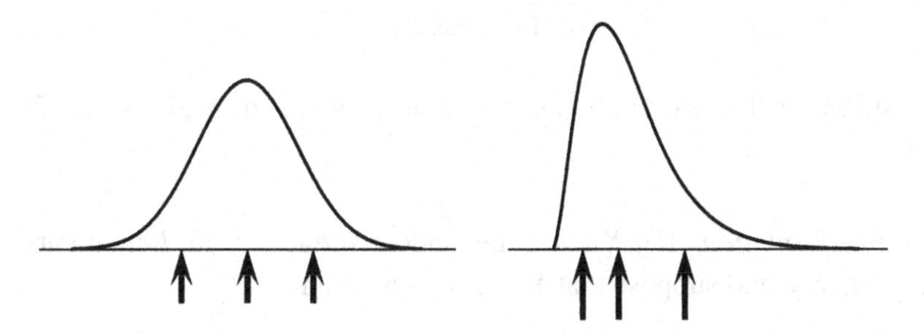

Fig. 2.3. Definition sketch for third L-moment.

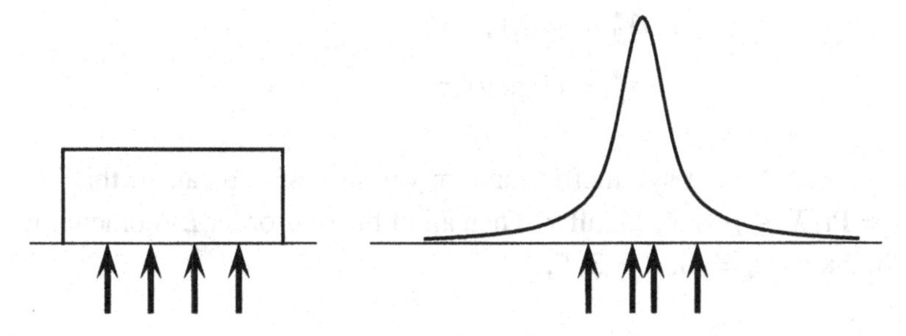

Fig. 2.4. Definition sketch for fourth L-moment.

Uniqueness. If the mean of the distribution exists, then the L-moments uniquely define the distribution, that is, no two distributions have the same L-moments.

Terminology

λ_1 is the L-location or mean of the distribution.
λ_2 is the L-scale.
τ is the L-CV.
τ_3 is the L-skewness.
τ_4 is the L-kurtosis.

Numerical values

λ_1 can take any value.
$\lambda_2 \geq 0$.
For a distribution that takes only positive values, $0 \leq \tau < 1$.

L-moment ratios satisfy $|\tau_r| < 1$ for all $r \geq 3$. Tighter bounds can be found for individual τ_r quantities. For example, bounds for τ_4 given τ_3 are

$$\tfrac{1}{4}(5\tau_3^2 - 1) \leq \tau_4 < 1\,. \tag{2.45}$$

For a distribution that takes only positive values, bounds for τ_3 given τ are $2\tau - 1 \leq \tau_3 < 1$.

Linear transformation. Let X and Y be random variables with L-moments λ_r and λ_r^*, respectively, and suppose that $Y = aX + b$. Then

$$\lambda_1^* = a\lambda_1 + b\,; \tag{2.46}$$

$$\lambda_2^* = |a|\lambda_2\,; \tag{2.47}$$

$$\tau_r^* = (\operatorname{sign} a)^r \tau_r\,, \qquad r \geq 3. \tag{2.48}$$

Symmetry. Let X be a symmetric random variable with mean μ, that is, $\Pr[X \geq \mu + x] = \Pr[X \leq \mu - x]$ for all x. Then all of the odd-order L-moment ratios of X are zero, that is, $\tau_r = 0$, $r = 3, 5, \ldots$.

L-moments have been calculated for many common distributions. A list is given in the appendix. Some cases are of particular interest. The uniform distribution has a quantile function $x(u)$ that is linear in u. We can show that a linear function $x(u)$ can be written as a weighted sum of $P_0^*(u)$ and $P_1^*(u)$. From the orthogonality relation,

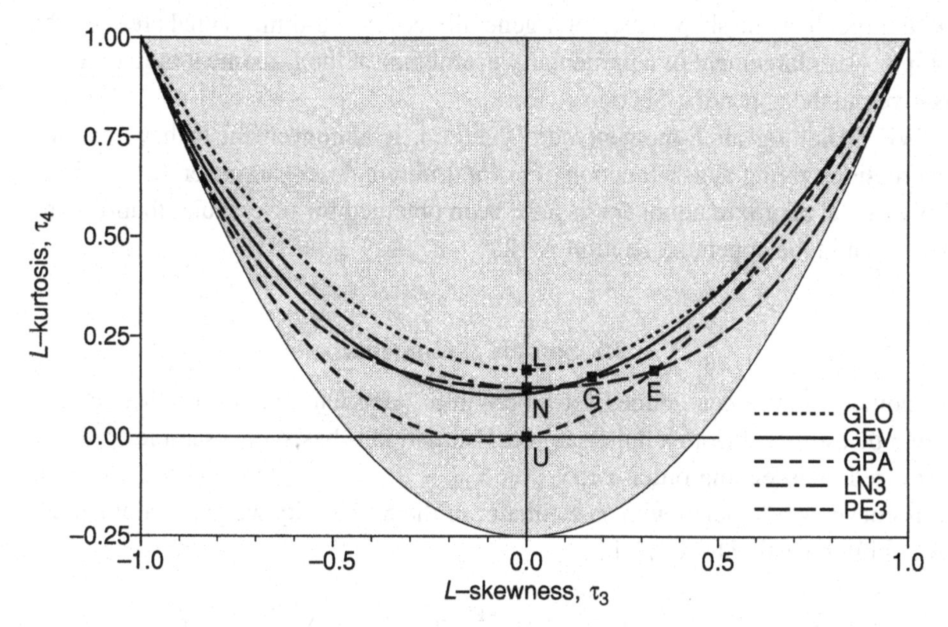

Fig. 2.5. L-moment ratio diagram. Two- and three-parameter distributions are shown as points and lines, respectively. Key to distributions: E – exponential, G – Gumbel, L – logistic, N – Normal, U – uniform, GPA – generalized Pareto, GEV – generalized extreme-value, GLO – generalized logistic, LN3 – lognormal, PE3 – Pearson type III. The shaded area contains the possible values of τ_3 and τ_4, given by (2.45).

(iii) in Section 2.4, it follows that $\int_0^1 x(u)P_r^*(u)\,du = 0$ for $r \geq 2$, whence all of the higher-order L-moments $\lambda_r, r \geq 3$, and the L-moment ratios $\tau_r, r \geq 3$, are zero for the uniform distribution. The distribution thus plays a central role in L-moment theory analogous to that of the Normal distribution in cumulant theory. The Normal distribution, being symmetric, has zero odd-order L-moments, but its even-order L-moments are not particularly simple. For example, the Normal distribution has $\tau_4 \approx 0.123$. The exponential distribution has particularly simple L-moment ratios: $\tau_3 = \frac{1}{3}$, $\tau_4 = \frac{1}{6}$.

A convenient way of representing the L-moments of different distributions is the L-moment ratio diagram, exemplified by Figure 2.5. This shows the L-moments on a graph whose axes are L-skewness and L-kurtosis. A two-parameter distribution with a location and a scale parameter plots as a single point on the diagram, because two distributions that differ only in their location and scale parameters are distributions of random variables X and $Y = aX + b$ with $a > 0$, and by Eq. (2.48) these random variables have the same L-skewness and L-kurtosis. A three-parameter distribution with location, scale, and shape parameters plots as a line, with different points on the line corresponding to different values of the shape parameter. Distributions with more than one shape parameter generally cover two-dimensional areas on the graph. An enlargement of a particularly useful part of the L-moment ratio diagram is given in the appendix, Section A.13.

When plotting an L-moment ratio diagram, it is convenient to have explicit expressions giving τ_4 as a function of τ_3 for different three-parameter distributions. Polynomial approximations for τ_4 have been obtained for several distributions and are given in the appendix, Section A.12.

2.7 Sample *L*-moments

L-moments have been defined for a probability distribution, but in practice must often be estimated from a finite sample. Estimation is based on a sample of size n, arranged in ascending order. Let $x_{1:n} \le x_{2:n} \le \ldots \le x_{n:n}$ be the ordered sample. It is convenient to begin with an estimator of the probability weighted moment β_r. An unbiased estimator of β_r is

$$b_r = n^{-1} \binom{n-1}{r}^{-1} \sum_{j=r+1}^{n} \binom{j-1}{r} x_{j:n} \tag{2.49}$$

(Landwehr et al., 1979a). This may alternatively be written as

$$b_0 = n^{-1} \sum_{j=1}^{n} x_{j:n} \,, \tag{2.50}$$

$$b_1 = n^{-1} \sum_{j=2}^{n} \frac{(j-1)}{(n-1)} x_{j:n} \,, \tag{2.51}$$

$$b_2 = n^{-1} \sum_{j=3}^{n} \frac{(j-1)(j-2)}{(n-1)(n-2)} x_{j:n} \,, \tag{2.52}$$

and in general

$$b_r = n^{-1} \sum_{j=r+1}^{n} \frac{(j-1)(j-2)\ldots(j-r)}{(n-1)(n-2)\ldots(n-r)} x_{j:n} \,. \tag{2.53}$$

Analogously to Eqs. (2.32)–(2.36), the *sample L-moments* are defined by

$$\ell_1 = b_0 \,, \tag{2.54}$$

$$\ell_2 = 2b_1 - b_0 \,, \tag{2.55}$$

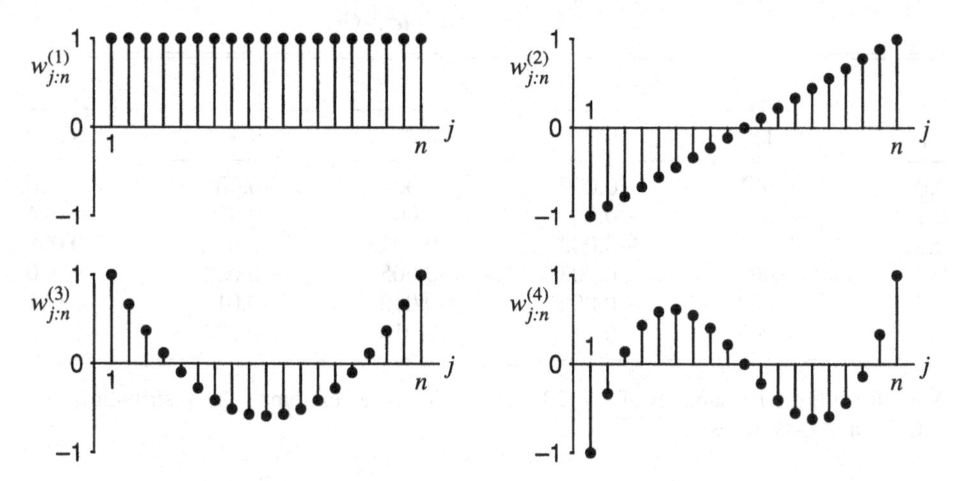

Fig. 2.6. Relative contributions of each observation to the first four sample *L*-moments, for sample size $n = 19$.

$$\ell_3 = 6b_2 - 6b_1 + b_0 \,, \tag{2.56}$$

$$\ell_4 = 20b_3 - 30b_2 + 12b_1 - b_0 \,, \tag{2.57}$$

and in general

$$\ell_{r+1} = \sum_{k=0}^{r} p^*_{r,k} b_k \,; \qquad r = 0, 1, \ldots, n-1; \tag{2.58}$$

the coefficients $p^*_{r,k}$ are defined as in Eq. (2.30). The sample *L*-moment ℓ_r is an unbiased estimator of λ_r.

From Eqs. (2.49) and (2.58), ℓ_r is a linear combination of the ordered sample values $x_{1:n}, \ldots, x_{n:n}$, and we can write

$$\ell_r = n^{-1} \sum_{j=1}^{n} w^{(r)}_{j:n} x_{j:n} \,. \tag{2.59}$$

The weights $w^{(r)}_{j:n}$ are illustrated in Figure 2.6 for the case $n = 19$; this shows the relative contributions of each observation to each sample *L*-moment. The weights have a pattern that resembles polynomials of degree $r - 1$ in j. Indeed, in the notation of Neuman and Schonbach (1974), $w^{(r)}_{j:n}$ is the discrete Legendre polynomial $(-1)^{r-1} P_{r-1}(j - 1, n - 1)$.

Table 2.1. *Bias of sample L-CV.*

	τ				
τ_3	0.1	0.2	0.3	0.4	0.5
0.0	−0.001	0.000	0.003	0.009	0.020
0.1	−0.001	−0.001	0.001	0.005	0.014
0.2	−0.001	−0.002	−0.001	0.001	0.008
0.3	−0.001	−0.003	−0.005	−0.004	0.000
0.4	−0.002	−0.006	−0.010	−0.012	−0.011
0.5	−0.003	−0.011	−0.018	−0.025	−0.027

Note: Results are for samples of size 20 from a generalized extreme value distribution with *L*-CV τ and *L*-skewness τ_3.

Analogously to Eqs. (2.37) and (2.38), the *sample L-moment ratios* are defined by

$$t_r = \ell_r/\ell_2 \tag{2.60}$$

and the *sample L-CV* by

$$t = \ell_2/\ell_1. \tag{2.61}$$

They are natural estimators of τ_r and τ, respectively.

The estimators t_r and t are not unbiased, but their biases are very small in moderate or large samples. Large-sample biases can be calculated using asymptotic theory (Hosking, 1990, p. 116). For example, the asymptotic bias of t_3 for a Gumbel distribution is $0.19n^{-1}$, and the asymptotic bias of t_4 for a Normal distribution is $0.03n^{-1}$, where n is the sample size.

Bias for smaller samples can be evaluated by simulation. It is generally the case that the bias of the sample *L*-CV, t, is negligible in samples of size 20 or more. For example, Table 2.1 gives the bias of t for samples of size 20 from a generalized extreme-value distribution. For a wide range of population τ and τ_3 values of the distribution, the bias of t is very small. Only when τ_3 exceeds 0.4, corresponding to a shape parameter $k < -0.33$, does the relative bias of t exceed 4%.

Figure 2.7 illustrates the bias of the sample *L*-skewness, t_3, and the sample *L*-kurtosis, t_4. Bias is shown graphically, by arrows that lead from the population values τ_3 and τ_4 to the means of the sample statistics t_3 and t_4. Results are based on 10,000 simulations of distributions with τ_3 and τ_4 at intervals of 0.05 over the range $0 \le \tau_3 \le 0.5$, $0 \le \tau_4 \le 0.3$. The parent distributions are kappa distributions when there exists a kappa distribution with the given τ_3 and τ_4 values

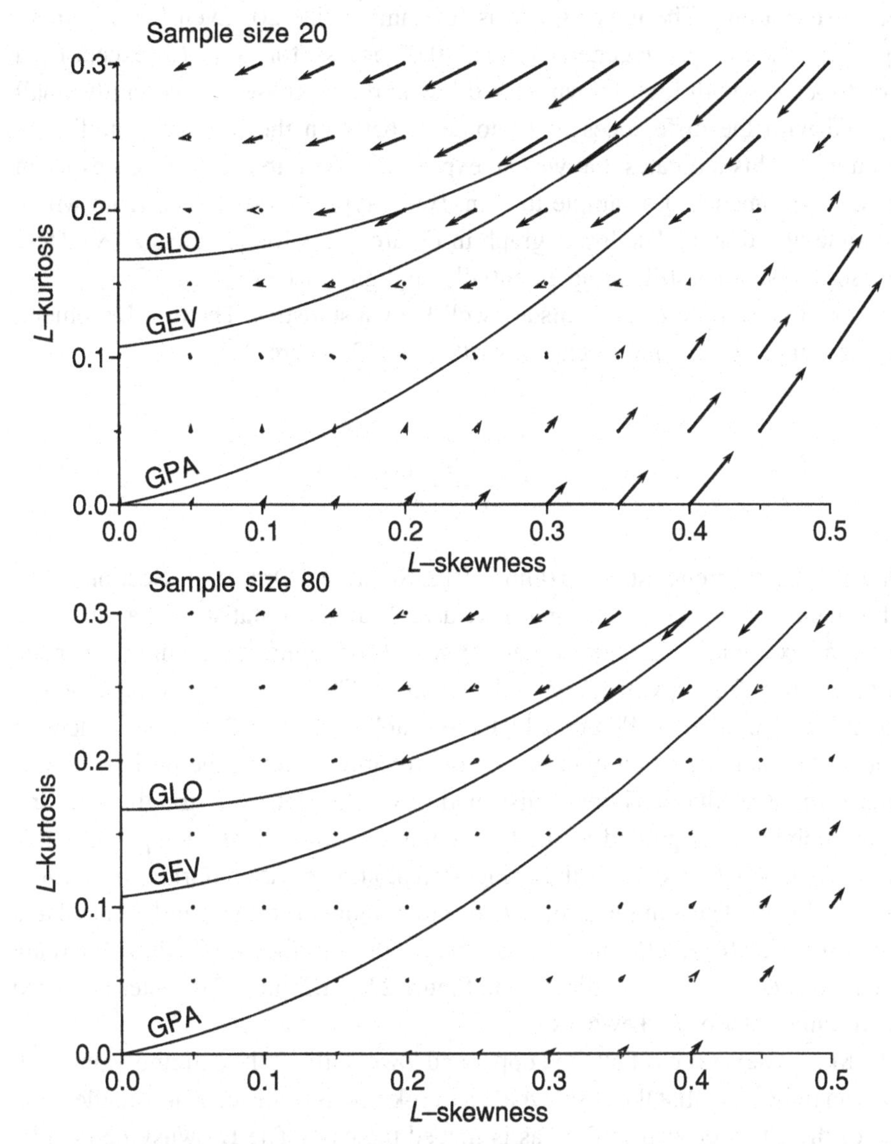

Fig. 2.7. Bias of sample L-skewness and L-kurtosis statistics. Upper graph is for sample size 20, lower graph is for sample size 80. Arrows lead from the population values to the mean of the sample statistics. Solid lines are the relationships between L-skewness and L-kurtosis for the generalized logistic (GLO), generalized extreme value (GEV), and generalized Pareto (GPA) distributions. Results are based on simulated samples drawn from kappa and Wakeby distributions.

(see Figure A.1 on page 204) and Wakeby distributions otherwise. Biases could of course be different for samples drawn from other parent distributions. The graphs also show the relationships between population L-skewness and population L-kurtosis for the generalized logistic, generalized extreme value, and generalized

Pareto distributions. The upper graph is for sample size 20. Even for this small sample size, the biases are generally small, 0.01 or less, for $\tau_4 < 0.2$, except for a few distributions with high skewness and low kurtosis. Biases are certainly small compared with the differences in L-moments between the different families of distributions. This indicates that we can expect to make unbiased inferences when using the L-moments of a sample to identify the type of distribution from which the sample was drawn. The lower graph in Figure 2.7 is for sample size 80. Here the biases are smaller still, being essentially negligible when $\tau_4 < 0.25$.

The first two sample L-moments are well-known statistics. The first L-moment, ℓ_1, is the sample mean, and ℓ_2 can be written as $\frac{1}{2}G$, where

$$G = \binom{n}{2}^{-1} \sum\sum_{1 \le i < j \le n} (x_{j:n} - x_{i:n}) \tag{2.62}$$

is Gini's mean difference statistic (Gini, 1912; Stuart and Ord, 1987, Section 2.22).

The quantities ℓ_1, ℓ_2 (or t), t_3, and t_4 are useful summary statistics of a sample of data. As an example we calculate them for six sets of annual maximum windspeed data taken from Simiu, Changery, and Filliben (1979). The data are tabulated in Table 2.2 and plotted in Figure 2.8. Each graph in Figure 2.8 is equivalent to a plot of the data values on extreme-value probability paper; the horizontal axis is transformed so that a Gumbel distribution would plot as a straight line, and the data point $x_{j:n}$ is plotted at the horizontal position $-\log(-\log p_{j:n})$, where $p_{j:n} = (j - 0.44)/(n + 0.12)$ is the Gringorten plotting position (Gringorten, 1963; Cunnane, 1978). The sample L-moments and L-moment ratios can be calculated using (2.49), (2.58), (2.60), and (2.61). The results are given in Table 2.3 and are plotted on an L-moment ratio diagram in Figure 2.9. In Table 2.3 the sites are listed in increasing order of L-skewness.

The Macon data set has flattened upper and lower tails on the extreme-value plot. This is an indication that the L-skewness and L-kurtosis of the data are smaller than those of the Gumbel distribution, as is indeed the case. The Brownsville data lie close to a straight line on Figure 2.8, reflecting the fact that its sample L-skewness and L-kurtosis, 0.1937 and 0.1509, are fairly close to the corresponding quantities, 0.1699 and 0.1504, for the Gumbel distribution. The Port Arthur data plot and L-moments are both similar to those of Brownsville. The Key West data set is more dispersed than the others; for Key West, the slope of the points on Figure 2.8 is greater and so is the L-scale statistic ℓ_2. The Corpus Christi data set contains a very high outlier, which accounts for its high L-skewness and L-kurtosis values. The Montgomery data plot, L-skewness and L-kurtosis are intermediate between those of Port Arthur and Corpus Christi.

Table 2.2. *Annual maximum windspeed data, in miles per hour, for six sites in the eastern United States.*

Macon, Ga., 1950–1977.									
32	32	34	37	37	38	40	40	40	42
42	42	43	44	45	45	46	48	49	50
51	51	51	53	53	58	58	60		
Brownsville, Tex., 1943–1977.									
32	33	34	34	35	36	37	37	38	38
39	39	40	40	41	41	42	42	43	43
43	44	44	46	46	48	48	49	51	53
53	53	56	63	66					
Port Arthur, Tex., 1953–1977.									
39	43	44	44	45	45	45	46	47	49
51	51	51	51	54	55	55	57	57	60
61	63	66	67	81					
Montgomery, Ala., 1950–1977.									
34	36	36	37	38	40	40	40	40	40
43	43	43	43	46	46	46	46	47	47
48	49	51	51	51	52	60	77		
Key West, Fla., 1958–1976.									
35	35	36	36	36	38	42	43	43	46
48	48	52	55	58	64	78	86	90	
Corpus Christi, Tex., 1943–1976.									
44	44	44	44	45	45	45	45	46	46
46	47	48	48	48	48	48	49	50	50
50	51	52	55	57	58	60	60	66	67
70	71	77	128						

2.8 Plotting-position estimators

A *plotting position* is a distribution-free estimator of $F(x_{j:n})$. Reasonable choices for plotting positions include $p_{j:n} = (j+\gamma)/(n+\delta)$ for $\delta > \gamma > -1$. Landwehr et al. (1979b) suggested the estimator

$$\tilde{\alpha}_r = n^{-1} \sum_{i=1}^{n} (1 - p_{j:n})^r \, x_{j:n} \tag{2.63}$$

of α_r. Analogously, we can define plotting-position estimators of λ_r and τ_r by

$$\tilde{\lambda}_r = n^{-1} \sum_{j=1}^{n} P^*_{r-1}(p_{j:n}) \, x_{j:n} \,, \tag{2.64}$$

$$\tilde{\tau}_r = \tilde{\lambda}_r / \tilde{\lambda}_2 \,. \tag{2.65}$$

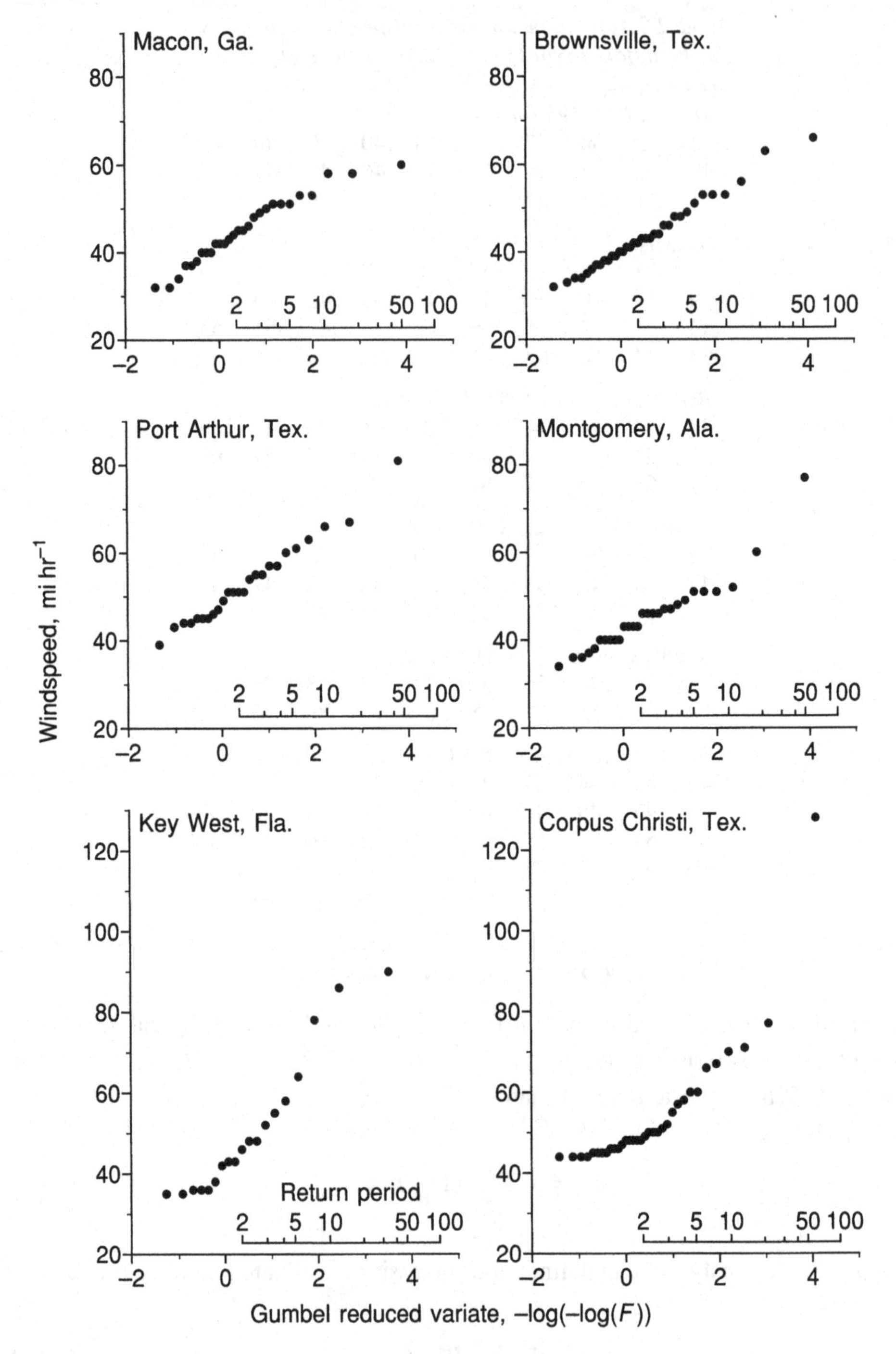

Fig. 2.8. Annual maximum windspeed data for the sites in Table 2.2.

Table 2.3. *L-moments of the annual maximum windspeed data in Table 2.2.*

Site	n	ℓ_1	ℓ_2	t	t_3	t_4
Macon	28	45.04	4.46	0.0990	0.0406	0.0838
Brownsville	35	43.63	4.49	0.1030	0.1937	0.1509
Port Arthur	25	53.08	5.25	0.0989	0.2086	0.1414
Montgomery	28	45.36	4.34	0.0958	0.2316	0.2490
Key West	19	51.00	9.29	0.1821	0.3472	0.1245
Corpus Christi	34	54.47	6.70	0.1229	0.5107	0.3150

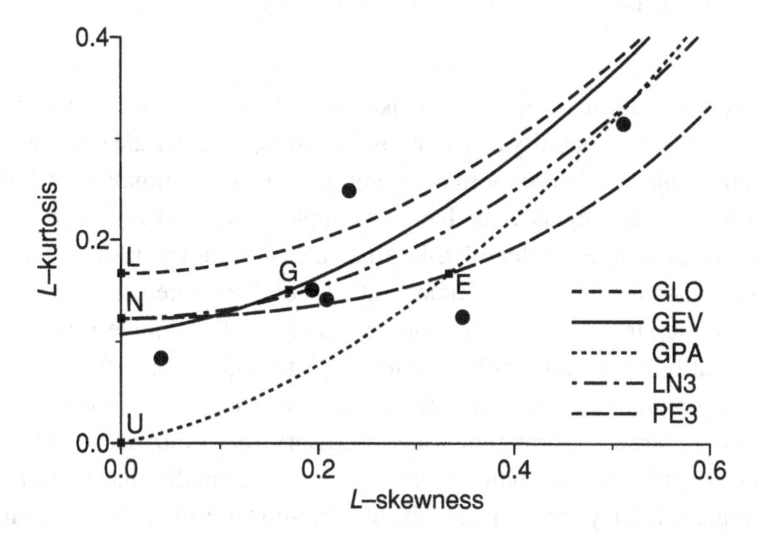

Fig. 2.9. *L*-moment ratio diagram for the annual maximum windspeed data in Table 2.2.

In general $\tilde{\lambda}_r$ is not an unbiased estimator of λ_r, but its bias tends to zero in large samples.

Plotting-position estimators were introduced by Landwehr et al. (1979b) for use in estimating the parameters of the Wakeby distribution. In particular, the choice $p_{j:n} = (j - 0.35)/n$ was found to give good results for the Wakeby and also for generalized extreme-value and generalized Pareto distributions (Hosking et al., 1985b; Hosking and Wallis, 1987a). These investigations were all primarily concerned with estimation of quantiles fitted to data with a physical lower bound of zero and *L*-CV and *L*-skewness both in the approximate range 0.1 to 0.3.

For other applications, these plotting-position estimators have significant disadvantages (Hosking and Wallis, 1995). Plotting-position estimators with $p_{j:n} = (j-0.35)/n$ are not invariant under location transformations of the data; if a constant is added to the data values, the estimators $\tilde{\lambda}_r$, $r \geq 2$, are changed. In extreme

Table 2.4. *Notation for moments and L-moments.*

	Population moment	Sample moment	Population *L*-moment	Sample *L*-moment
Location (mean)	μ	$\bar{x}$	λ_1	ℓ_1
Scale	σ	s	λ_2	ℓ_2
CV	C_v	$\hat{C}_v$	τ	t
Skewness	γ	g	τ_3	t_3
Kurtosis	κ	k	τ_4	t_4

Note: In other chapters γ is used to denote Euler's constant, 0.5772..., and both γ and k are used to denote parameters of certain probability distributions.

cases, plotting-position estimators can take values that would be impossible for the *L*-moments of any probability distribution; for example, the scale estimator $\tilde{\lambda}_2$ can take negative values. Most importantly, plotting-position estimators of *L*-moment ratios have generally higher bias than the sample *L*-moment ratios. In particular, the plotting-position estimator of *L*-kurtosis has substantial positive bias. When estimation of *L*-moments and *L*-moment ratios is of direct interest, rather than as an intermediate step towards the estimation of quantiles, the sample *L*-moments and *L*-moment ratios are greatly preferable to the plotting-position estimators.

In this monograph we work throughout with the estimators ℓ_r, t and t_r. We refer to them as "unbiased" estimators when necessary for comparison with plotting-position estimators; the quotation marks serve as a reminder that t and t_r are not exactly unbiased. They are inferior to the plotting-position estimators only for some instances of estimation of extreme quantiles in regional frequency analysis and have generally lower bias as estimators of the *L*-moment ratios, τ_r. This makes them more suitable than the plotting-position estimators for the applications in Chapters 3–5, which involve using the *L*-moments of a data sample to summarize the properties of the sample and to infer the shape of the underlying population from which the sample was drawn.

2.9 Moments and *L*-moments

Both moments and *L*-moments are measures of the location, scale, and shape of probability distributions. Here we consider their similarities and differences. For reference, Table 2.4 gives our notation for moments and *L*-moments. First we compare the individual moment and *L*-moment quantities.

The *L*-moment measure of location is the mean, λ_1. This is, of course, the same as the first moment μ.

To compare λ_2 with the moment-based scale measure σ, the standard deviation, we write

$$\lambda_2 = \tfrac{1}{2}\,\mathrm{E}(X_{2:2} - X_{1:2}), \qquad \sigma^2 = \tfrac{1}{2}\,\mathrm{E}(X_{2:2} - X_{1:2})^2. \tag{2.66}$$

Both quantities measure the difference between two randomly drawn elements of a distribution, but σ^2 gives relatively more weight to the largest differences.

The two quantities satisfy the inequality

$$\sigma \geq \sqrt{3}\,\lambda_2\,, \tag{2.67}$$

which follows from Plackett (1947); equality is achieved only by the uniform distribution. Many moderately skew distributions have $\sigma \approx 2\lambda_2$; the exact equality $\sigma = 2\lambda_2$ holds for the exponential distribution ($\tau_3 = 0.3333$) and for a generalized extreme-value distribution with $\tau_3 = 0.2628$.

CV and L-CV are related similarly to σ and λ_2. Their estimators satisfy

$$\hat{C}_v \geq \left(\frac{3n}{n+1}\right)^{1/2} t\,, \tag{2.68}$$

which follows from Barker (1983). This bound is almost reached by many samples from symmetric or nearly symmetric distributions. For example, samples of size 50 from the Normal distribution lie close to this bound, as illustrated by the left graph of Figure 2.10. For moderately skew distributions $\hat{C}_v$ is often approximately twice as large as t, but if outliers are present $\hat{C}_v$ is larger still. See, for example, the right graph of Figure 2.10.

Figure 2.11 compares the skewness measures τ_3 and γ for different distributions. For symmetric distributions, both τ_3 and γ are zero, and many near-symmetric distributions have $\gamma \approx 6\tau_3$, but in general there is no simple relationship between γ and τ_3. Both γ and τ_3 may yield a large positive skewness either when a distribution has a heavy right tail or when a continuous distribution is reverse J-shaped, that is, has a finite lower bound near which $f(x) \to \infty$. The former case tends to yield particularly high values of γ relative to τ_3, because γ is more sensitive to the extreme tail weight of the distribution. Indeed for some heavy-tailed distributions, γ approaches infinity while τ_3 has still quite a modest value, for example, 0.33 in the case of the generalized logistic distribution.

Kurtosis, as measured by the moment ratio κ, has no unique interpretation. It can be thought of as the "peakedness" of a distribution, or as "tail weight," but only for fairly closely defined families of symmetric unimodal distributions do these interpretations have any demonstrable validity (see Balanda and MacGillivray,

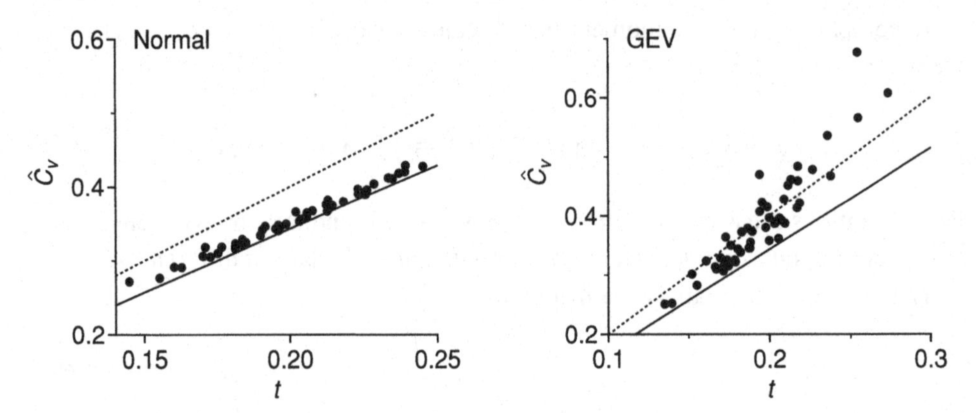

Fig. 2.10. Sample CV, $\hat{C}_v$, and sample *L*-CV, t, for 50 samples of size 50 simulated from a Normal distribution with $\tau = 0.2$ (left graph) and a generalized extreme-value distribution with $\tau = 0.2$ and $\tau_3 = 0.3$ (right graph). The solid line on each graph is the lower bound (2.68); the dotted line is the reference line $\hat{C}_v = 2t$.

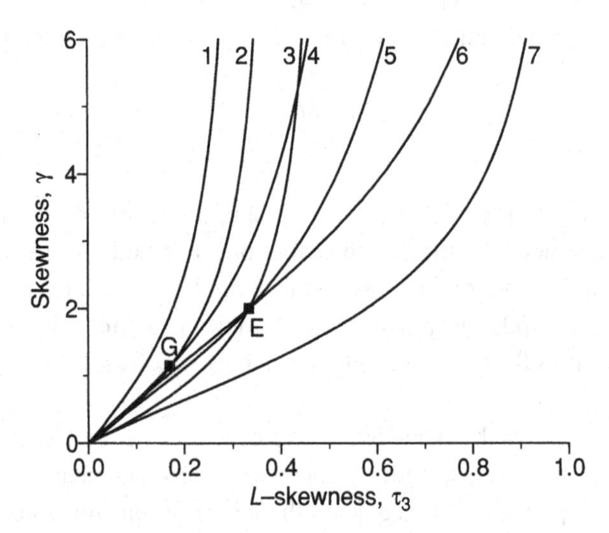

Fig. 2.11. Comparison of skewness and *L*-skewness. Key to distributions: E – exponential, G – Gumbel, 1 – generalized logistic, 2 – generalized extreme value, 3 – generalized Pareto, 4 – lognormal, 5 – Pearson type III, 6 – Weibull (reverse generalized extreme value), 7 – reverse generalized Pareto.

1988, and references therein). *L*-kurtosis, τ_4, is equally difficult to interpret uniquely and is best thought of as a measure similar to κ but giving less weight to the extreme tails of the distribution.

Both sample *L*-skewness and sample *L*-kurtosis are much less biased than the ordinary skewness and kurtosis. Figure 2.7, for example, shows much lower biases

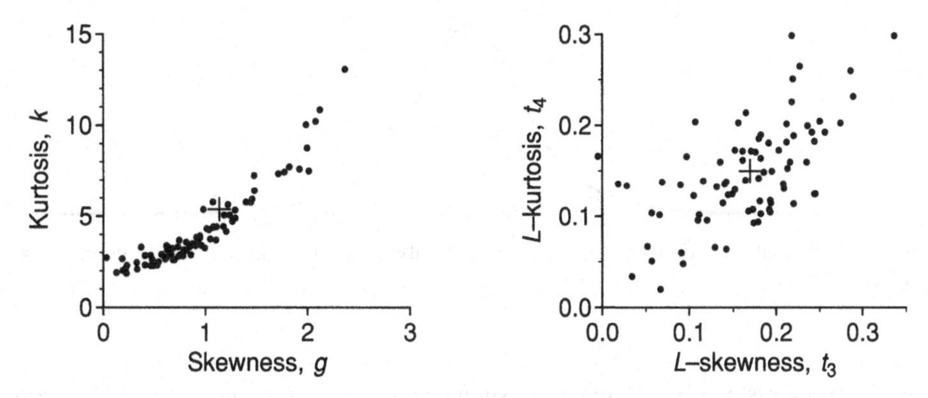

Fig. 2.12. Sample moment ratios and *L*-moment ratios for 80 samples of size 50 simulated from a Gumbel distribution. The + marks indicate the population moment ratios and *L*-moment ratios for the Gumbel distribution.

than those found for the conventional skewness statistic, g, by Wallis et al. (1974) for similar distributions. Royston (1992, Figure 7) illustrates similar results for the lognormal distribution.

The joint distribution of the ordinary sample skewness and sample kurtosis is asymptotically Normal, but in small and moderate samples the asymptotic distribution is a poor approximation, particularly when the underlying distribution is even moderately skew. Figure 2.12 shows the joint distribution of sample moment ratios and *L*-moment ratios for 80 random samples of size 50 simulated from a Gumbel distribution. The sample skewness and kurtosis values lie close to a single line in the (g, k) plane, indicating that k gives little information about the sample additional to that conveyed by g. The joint distribution of g and k is clearly far from Normal. The sample *L*-skewness and *L*-kurtosis, in contrast, have a joint distribution that is nondegenerate and appears to be close to bivariate Normal. This near-Normality of the sampling distributions is a property that we make use of in some of the procedures described in later chapters.

Now we consider more general properties of moments and *L*-moments.

As mentioned earlier, *L*-moments exist whenever the mean of the distribution exists. This includes cases in which some of the higher moments fail to exist. These cases do occur in practice. For example, for the generalized extreme-value distribution the third and fourth moments fail to exist when the distribution's shape parameter k satisfies $k \leq -\frac{1}{3}$ and $k \leq -\frac{1}{4}$, respectively. At these k values the *L*-moment ratios take the fairly moderate values $\tau_3 = 0.403$ and $\tau_4 = 0.241$, respectively. Data samples that yield sample *L*-moment ratios as large as this occur frequently in the analysis of some kinds of data (e.g., windspeed data such as those in Table 2.2 or the annual maximum streamflow data of Section 9.2). Although

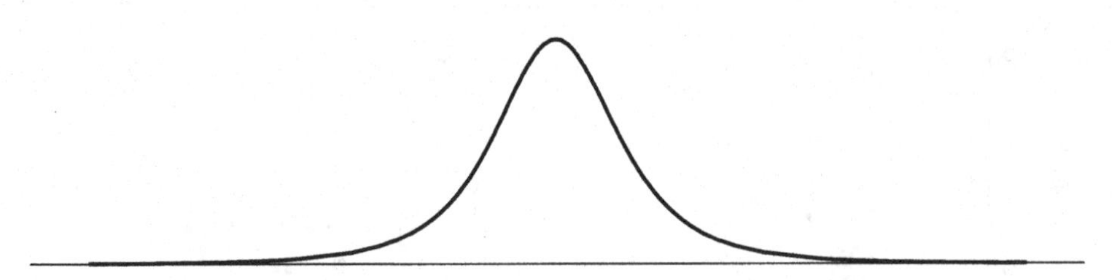

Fig. 2.13. Probability density function of a lambda distribution with parameter $\lambda = -0.1466$.

sample moments can be calculated in such cases, they cannot be expected to yield reliable information about the form of the underlying distribution.

Moment ratios are unbounded, whereas *L*-moment ratios have a natural bound $|\tau_r| < 1$. We consider the boundedness of *L*-moment ratios to be an advantage. Intuitively, it seems easier to interpret a measure such as τ_3, which is constrained to lie within the interval $-1 < \tau_3 < 1$, than the ordinary skewness, which can take arbitrarily large positive or negative values.

Algebraic bounds on the sample moment ratios were mentioned in Section 2.3. *L*-moment ratios are not subject to such restrictive bounds. The sample *L*-moment ratios can take any values that the population *L*-moment ratios can (Hosking, 1990).

Perhaps the main difference between moments and *L*-moments is that moments give greater weight to the extreme tails of the distribution. This can be seen by comparing Eqs. (2.28) and (2.31). As r increases, the weight given to the tail of the distribution, $u \approx 1$, increases as $\{x(u)\}^r$ in Eq. (2.28) but as u^r in Eq. (2.31). For most distributions, $x(u)$ increases much faster than u as u approaches 1; for distributions with no upper bounds, of course, $x(u) \to \infty$ as $u \to 1$.

To illustrate the greater dependence of the ordinary moments on the extreme tails of a distribution, consider a lambda distribution (Tukey, 1960) with quantile function $x(u) = (u^\lambda - (1-u)^\lambda)/\lambda$ and $\lambda = -0.1466$. If the distribution is truncated at its 0.001 and 0.999 quantiles, its kurtosis κ falls from 10.00 to 5.48, but its *L*-kurtosis τ_4 falls only from 0.224 to 0.204. The probability density function of the truncated distribution is shown in Figure 2.13. On the scale of the figure, the tails of the distribution beyond the truncation points are indistinguishable from zero.

Sample moments, too, are more affected than their *L*-moment analogs by extreme observations. For example, consider the Corpus Christi windspeed data given in Table 2.2 and Figure 2.8. If the largest observation is deleted, the sample coefficient of variation $\hat{C}_v$ falls from 0.289 to 0.173 and the sample skewness g falls from 3.37 to 1.32, falls of 40% and 61%, respectively. The sample *L*-CV t falls from 0.1229 to 0.0908 and the sample *L*-skewness t_3 falls from 0.5107 to 0.3721, falls of only 26% and 27%, respectively. Vogel and Fennessey (1993) show that, even for

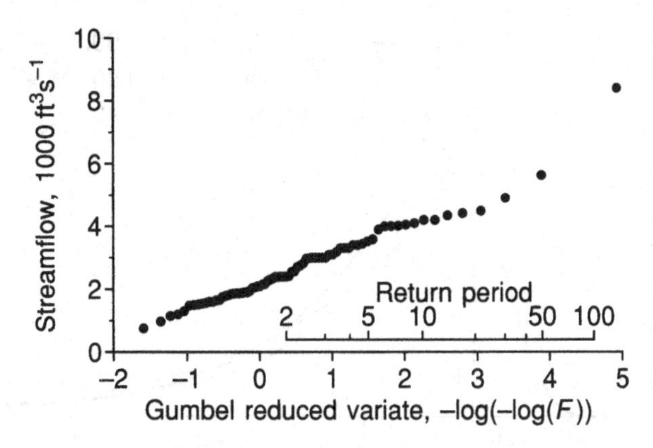

Fig. 2.14. Annual maximum streamflows, Oconto River near Gillett, Wis.

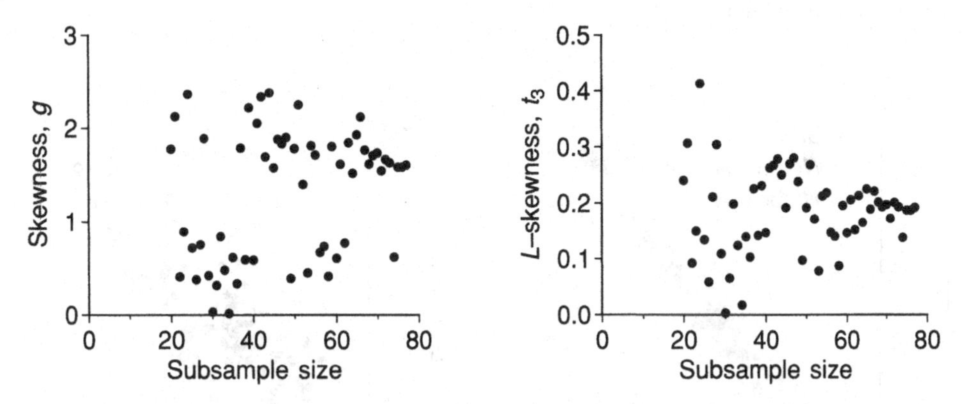

Fig. 2.15. Skewness and L-skewness of subsamples of the Oconto River data of Fig. 2.14.

sample sizes in excess of 5000, g can be severely affected by an outlier, whereas t_3 is not.

As a further example, consider the data set shown in Figure 2.14. The data, obtained from the U.S. Geological Survey, are annual maximum streamflows at USGS site 04071000, Oconto River near Gillett, Wis., for the period 1907–1987. There are four missing values, so the sample size is 77. The largest data value is to some extent an outlier, being nearly 50% larger than the second-largest value. Subsamples of the data were obtained by randomly deleting data points, and moments and L-moments were calculated for the subsamples. The results for subsamples of size 20, 21, ..., 77 are shown in Figure 2.15. The ordinary sample skewness appears to take two distinct values depending on whether the outlier is or is not in the subsample. In contrast, the sample L-skewness, though clearly affected by the outlier and quite variable in small samples, is fairly stable around its

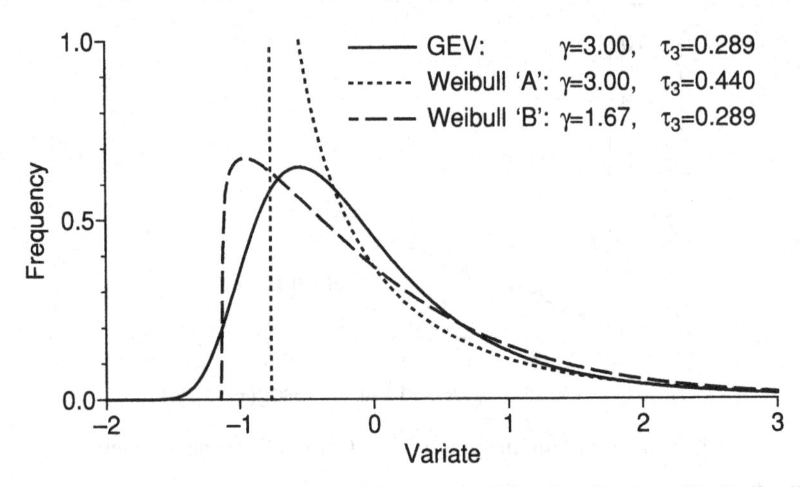

Fig. 2.16. Probability density functions of three probability distributions. Each distribution has zero mean and unit variance.

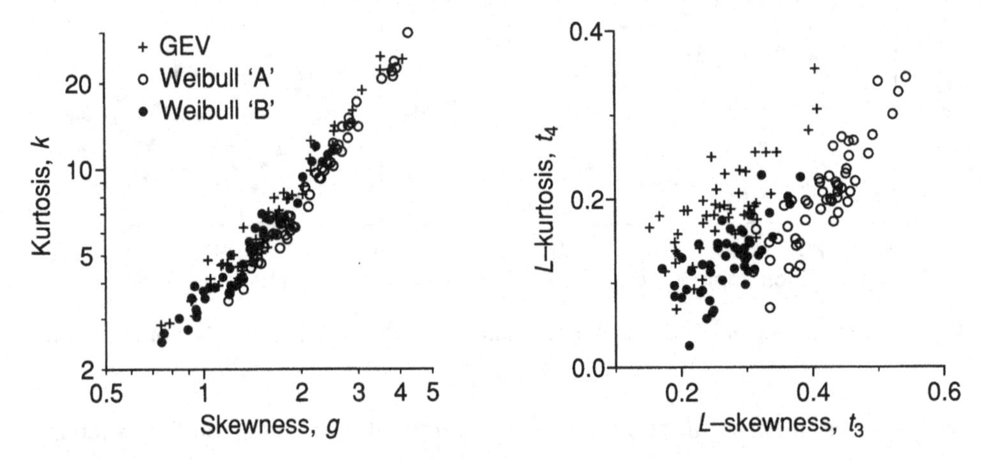

Fig. 2.17. Sample skewness versus sample kurtosis and sample L-skewness versus sample L-kurtosis, for samples of size 100 simulated from the three distributions of Fig. 2.16.

complete-sample value of 0.1918. This presentation follows Royston (1992), who gave a similar example for medical data.

An important application of summary statistics calculated from an observed random sample is identification of the distribution from which the sample was drawn. This is much more easily achieved, particularly for skew distributions, by using L-moments rather than conventional moments. The following example is from Hosking (1990). Using Monte Carlo simulation, 50 random samples of size 100 were generated from each of three distributions: a generalized extreme-

Table 2.5. *Estimated parameters and quantiles for GEV distributions fitted to the windspeed data in Table 2.2.*

	Parameters			Quantiles, $x(F)$				
				F				
Site	ξ	α	k	.01	.10	.50	.90	.99
Macon	42.1	7.56	0.212	28	35	45	56	64
Brownsville	39.8	6.26	−0.037	30	35	42	55	71
Port Arthur	48.5	7.15	−0.059	38	43	51	66	86
Montgomery	41.5	5.70	−0.094	33	37	44	56	74
Key West	41.9	9.89	−0.258	30	35	46	72	129
Corpus Christi	47.5	4.87	−0.471	42	44	49	67	127

value (GEV) distribution with skewness 3.0 and two Weibull distributions, one with the same skewness and one with the same *L*-skewness as this GEV distribution. The distributions are illustrated in Figure 2.16. Moments and *L*-moments of the generated samples are shown in Figure 2.17. The sample conventional moments from the three distributions all lie close to a single line on the graph and overlap each other; they offer little hope of identifying the population distribution. In contrast, the sample *L*-moments plot as fairly well separated groups and permit a high probability of discrimination between the three distributions.

2.10 Parameter estimation using *L*-moments

A common problem in statistics is the estimation, from a random sample of size n, of a probability distribution whose specification involves a finite number, p, of unknown parameters. Analogously to the usual method of moments, the *method of L-moments* obtains parameter estimates by equating the first p sample *L*-moments to the corresponding population quantities. This requires expressions for the parameters in terms of the *L*-moments. Such expressions have been obtained for many standard distributions. Examples are given in the appendix.

As an example, we fit the generalized extreme-value distribution to the six windspeed data sets given in Table 2.2. The sample *L*-moments are given in Table 2.3. Parameter estimates are calculated by substituting the sample *L*-moments into Eqs. (A.55) and (A.56). Quantile estimates are obtained by substituting the estimated parameters into Eq. (A.44). Results are given in Table 2.5. The fitted distributions are graphed in Figure 2.18.

Exact distributions of parameter estimators obtained by the method of *L*-moments are in general difficult to derive, but large-sample approximations can be obtained

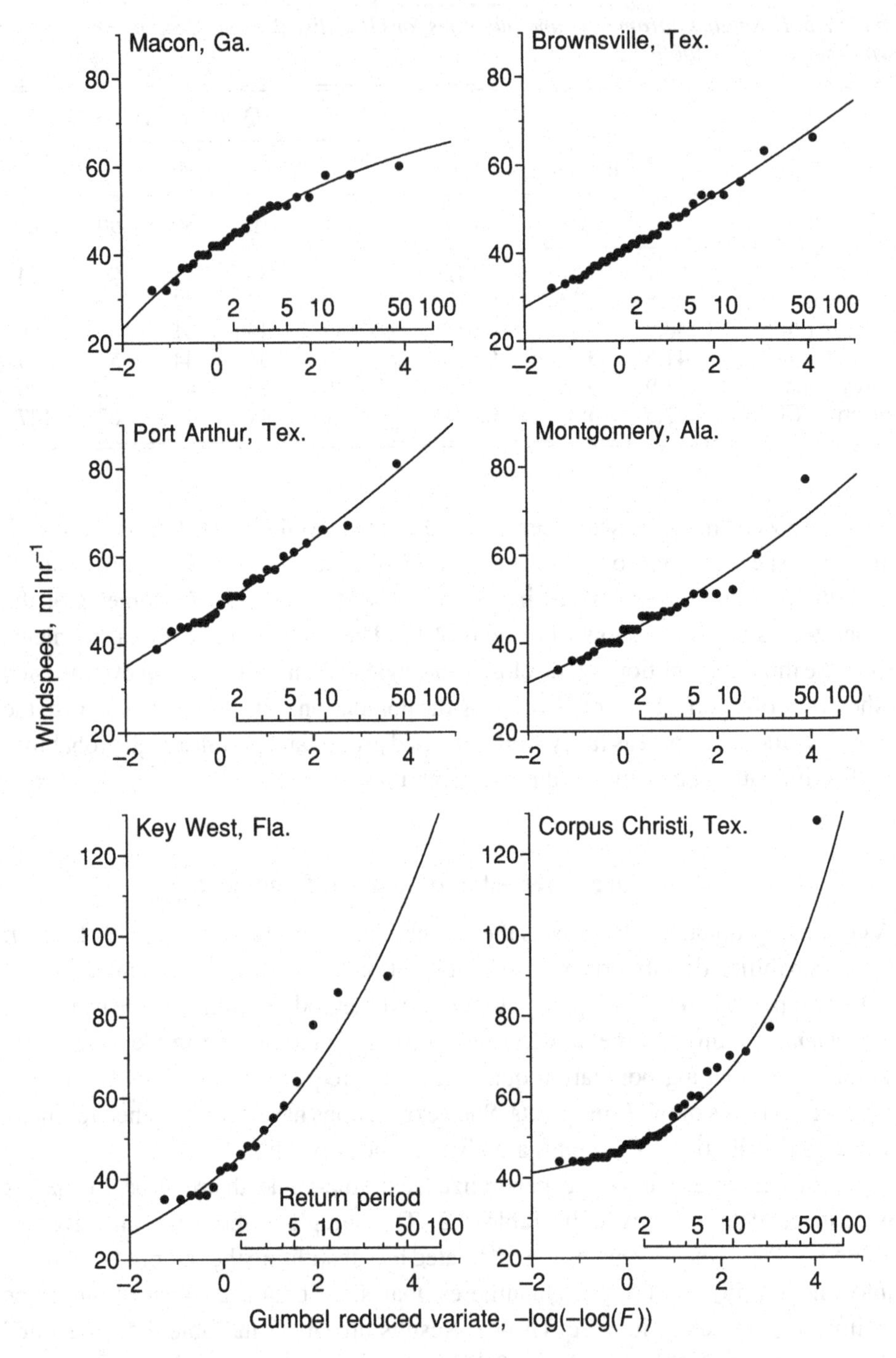

Fig. 2.18. GEV distributions fitted to the windspeed data of Table 2.2.

by asymptotic theory. Hosking (1986a) gives several examples of such results. For most standard distributions, this approach can be used to show that L-moment estimators of parameters and quantiles are asymptotically Normally distributed and also to find standard errors and confidence intervals.

Hosking et al. (1985b) and Hosking and Wallis (1987a) found that with small and moderate samples the method of L-moments is often more efficient than maximum likelihood. These results are for estimators based on a single sample of data, and are not directly relevant to regional frequency analysis. However, they demonstrate that the method of L-moments yields efficient and computationally convenient estimates of parameters and quantiles, and we may reasonably expect these properties to continue to hold in index-flood procedures for regional frequency analysis.

3

Screening the data

3.1 The importance of screening the data

The first essential of any statistical data analysis is to check that the data are appropriate for the analysis. For frequency analysis, the data collected at a site must be a true representation of the quantity being measured and must all be drawn from the same frequency distribution. An initial screening of the data should aim to verify that these requirements are satisfied.

The exact nature of the problems that may affect the data depend on the kind of data that were measured. For environmental data for which a frequency analysis is being attempted, two kinds of error are particularly important and plausible.

First, data values may be incorrect. Incorrect recording or transcription of data values is easily done and casts doubt on any subsequent frequency analysis of the data.

Second, the circumstances under which the data were collected may have changed over time. The measuring device may have been moved to a different location or trends over time may have arisen from changes in the environment of the measuring device. This means that the frequency distribution from which the data were sampled is not constant over time, and frequency analysis of the data will not be a valid basis for estimating the probability distribution of future measurements at the site.

Even though the data may reputedly be reliable, it is still important to check for errors. A sobering example was provided by Wallis, Lettenmaier, and Wood (1991), who compiled a set of daily precipitation and temperature records for 1009 sites in the United States from data supplied by the National Climatic Data Center (NCDC). The data had already been collected by NCDC from the original sources and had undergone some validity checking and preprocessing for incorporation into NCDC's Historical Climatology Network. However, 38% of the sites had at least one occurrence of such gross errors as daily precipitation less than zero or more than 20 in (500 mm), daily temperature range (maximum minus minimum) less than zero or more than 100°F (56°C), or maximum temperature less than 20°F (−7°C) during the period May–September.

In regional frequency analysis, related data are available for several or many sites. At least three kinds of checks on the data can be useful. Checks of individual data values can reveal the kind of gross errors found by Wallis et al. (1991). Checks of each site's data separately can identify outlying values and repeated values, which may arise from errors in recording or transcribing the data. Checks for trends and changes in level of the data are also useful. Comparisons between data from different sites can reveal many kinds of data irregularities. If a site is discordant with other apparently similar sites, then there may be a problem with the data for the discordant site.

Tests for outliers and trends are well established in the statistical literature (e.g., Barnett and Lewis, 1994; Kendall, 1975). For comparison of data from different sites, some techniques, such as double-mass plots or quantile–quantile plots, are well known. In the context of regional frequency analysis using *L*-moments, we have found that useful information can be obtained by comparing the sample *L*-moment ratios for different sites. Incorrect data values, outliers, trends, and shifts in the mean of a sample can all be reflected in the *L*-moments of the sample. A convenient amalgamation of the *L*-moment ratios into a single statistic, a measure of the discordancy between the *L*-moment ratios of a site and the average *L*-moment ratios of a group of similar sites, is presented in Section 3.2.

3.2 A discordancy measure

3.2.1 Aim

Given a group of sites, the aim is to identify those sites that are grossly discordant with the group as a whole. Discordancy is measured in terms of the *L*-moments of the sites' data.

3.2.2 Heuristic description

Regard the sample *L*-moment ratios (*L*-CV, *L*-skewness, *L*-kurtosis) of a site as a point in three-dimensional space. A group of sites will yield a cloud of such points. Flag as discordant any point that is far from the center of the cloud. "Far" is interpreted in such a way as to allow for correlation between the sample *L*-moment ratios.

For example, see Figure 3.1. For convenience we consider *L*-CV and *L*-skewness only. The center of the cloud of points, marked by +, is the point whose coordinates are the group average values of *L*-CV and *L*-skewness. We construct concentric ellipses with major and minor axes chosen to give the best fit to the data, as determined

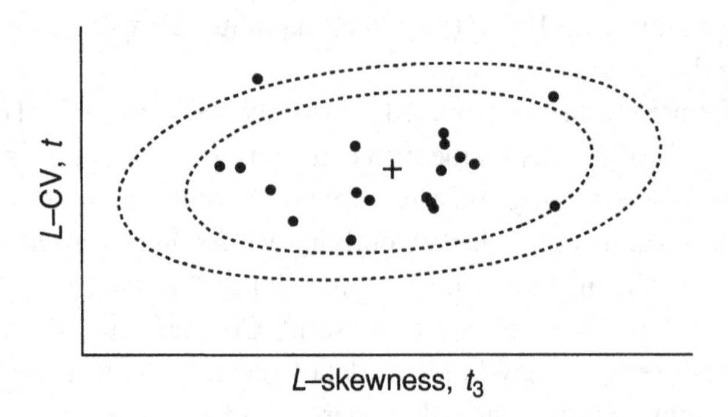

Fig. 3.1. Definition sketch for discordancy.

by the sample covariance matrix of the sites' *L*-moment ratios. "Discordant" points are those that lie outside the outermost ellipse.

3.2.3 Formal definition

Suppose that there are N sites in the group. Let $\mathbf{u}_i = [t^{(i)} \quad t_3^{(i)} \quad t_4^{(i)}]^{\mathrm{T}}$ be a vector containing the t, t_3, and t_4 values for site i: the superscript T denotes transposition of a vector or matrix. Let

$$\bar{\mathbf{u}} = N^{-1} \sum_{i=1}^{N} \mathbf{u}_i \tag{3.1}$$

be the (unweighted) group average. Define the matrix of sums of squares and cross-products,

$$\mathbf{A} = \sum_{i=1}^{N} (\mathbf{u}_i - \bar{\mathbf{u}})(\mathbf{u}_i - \bar{\mathbf{u}})^{\mathrm{T}}. \tag{3.2}$$

Define the discordancy measure for site i,

$$D_i = \tfrac{1}{3} N (\mathbf{u}_i - \bar{\mathbf{u}})^{\mathrm{T}} \mathbf{A}^{-1} (\mathbf{u}_i - \bar{\mathbf{u}}). \tag{3.3}$$

Declare site i to be discordant if D_i is large. The definition of "large" depends on the number of sites in the group. We suggest that a site be regarded as discordant if its D_i value exceeds the critical value given in Table 3.1.

Table 3.1. *Critical values for the discordancy statistic D_i.*

Number of sites in region	Critical value	Number of sites in region	Critical value
5	1.333	10	2.491
6	1.648	11	2.632
7	1.917	12	2.757
8	2.140	13	2.869
9	2.329	14	2.971
		≥ 15	3

3.2.4 Notes

D_i is a standard discordancy measure for multivariate observations. Wilks (1963) proposed an outlier measure that is equivalent to the largest of the D_i. For a univariate observation, D_i reduces to a multiple of $(u_i - \bar{u})^2/s^2$, the squared studentized residual; the maximum absolute studentized residual has been widely used as an outlier measure since its introduction by Thompson's (1935) work. The average of D_i over all sites is 1.

It is not easy to choose a value of D_i that can be used as a criterion for deciding whether a site is discordant. Hosking and Wallis (1993) initially suggested the criterion $D_i \geq 3$, but this is not satisfactory for small regions. We can show that D_i satisfies the algebraic bound

$$D_i \leq (N - 1)/3; \tag{3.4}$$

thus, for example, values of D_i larger than 3 can occur only in regions having 11 or more sites. To some extent, the criterion for discordancy should be an increasing function of the number of sites in the region. This is because large regions are more likely to contain sites with large values of D_i. However, we still recommend that any site with $D_i > 3$ be regarded as discordant, as such sites have *L*-moment ratios that are markedly different from the average for the other sites in the region.

A discordancy criterion for small regions can be derived from theoretical considerations. If it is assumed that the $\mathbf{u}_i$ are drawn from independent identical multivariate Normal distributions, then the distribution of Wilks's statistic can be derived (see Wilks, 1963, or Caroni and Prescott, 1992), and a significance test for the presence of an outlier can be obtained. For a test with significance level α an approximate critical value of $\max_i D_i$ is $(N - 1)Z/(N - 4 + 3Z)$, where Z is the upper $100\alpha/N$ percentage point of an F distribution with 3 and $N - 4$ degrees of freedom. This critical value is a function of α and N. We regard a choice of $\alpha = 0.10$ as reasonable; with this choice of α there is approximately a 10% chance that in a group of sites that in fact constitute a homogeneous region, at least one site would be flagged as discordant. For $\alpha = 0.10$ and $5 \le N < 15$ the critical values are those tabulated in Table 3.1. For $N \ge 15$ the critical value is greater than 3, but, as noted above, we recommend that sites with $D_i > 3$ be regarded as discordant anyway. In any case, it is advisable to examine the data for the sites with the largest D_i values, regardless of the magnitude of these values.

In very small regions the statistic D_i is not very informative. When $N \le 3$ the matrix $\mathbf{A}$ is singular and D_i cannot be calculated. When $N = 4$ each D_i value is 1. When $N = 5$ or $N = 6$ the critical values in Table 3.1 are very close to the bound (3.4). Thus D_i is likely to be useful only for regions with $N \ge 7$. This should not be a complete surprise, as it is difficult to judge whether a site is grossly unusual when there are few other sites with which to compare it.

The use of an unweighted average in the definition of $\bar{\mathbf{u}}$ is preferred to the weighted average used in the heterogeneity and goodness-of-fit measures described in Chapters 4 and 5. A weighted average allows for greater variability in short records and would permit a short-record site to be further from the group average before being flagged as discordant. However, the D_i statistics are calculated at an early stage of the analysis, when it is important to identify the unusual sites and their potential data errors regardless of the record length.

3.3 Use of the discordancy measure

Two uses for the discordancy measure are envisaged.

First, at the outset of the analysis the discordancy measure may be applied to a large group of sites, all those within some large geographical area. The idea is that sites with gross errors in their data will stand out from the other sites and be flagged as discordant. Sites flagged as discordant at this stage should therefore be closely scrutinized for errors in the recording or transcription of data or for sources of unreliability in the data, such as a recording gage having been moved, or for man-induced changes of the site's frequency distribution over time.

Later in the analysis, when homogeneous regions have been at least tentatively identified, the discordancy measure can be calculated for each site in a proposed region. If any site is discordant with the region as a whole, the possibility of moving that site to another region should be considered. It must be borne in mind, however, that a site's L-moments may differ by chance alone from those of other physically similar sites. For example, an extreme but localized meteorological event may have affected only a few sites in a region. If such an event is approximately equally likely to affect any of the sites in the future, then it is correct to treat the entire group of sites as a homogeneous region, even though some sites may appear to be discordant with the region as a whole.

Table 3.2. *Summary statistics for annual maximum streamflow data for sites in USGS Region 3 with small drainage basin area.*

HUC	Area	n	ℓ_1	t	t_3	t_4	D_i
02099000	14	60	2010	0.3287	0.2353	0.1699	0.62
02111000	28	48	1979	0.5003	0.5995	0.4781	2.22
02044000	38	42	2520	0.3541	0.3787	0.3702	0.80
02051000	55	40	3378	0.3094	0.3937	0.4542	2.69
02267000	58	38	90	0.2400	0.3447	0.2349	3.88*
02138500	66	66	5428	0.3979	0.4127	0.3110	0.11
02310000	72	44	1209	0.4396	0.3595	0.2111	0.91
02088000	83	48	1735	0.3716	0.2893	0.1900	0.40
02143000	83	52	5687	0.3317	0.2596	0.1135	0.90
02111500	89	48	4487	0.3623	0.3755	0.2883	0.05
02065500	98	42	2467	0.4234	0.4064	0.2979	0.29
02070000	108	60	4032	0.4313	0.4883	0.2993	1.16
02046000	112	42	2754	0.4400	0.4039	0.2140	0.90
02074500	112	59	4775	0.3803	0.4176	0.2684	0.53
02154500	116	58	3549	0.3250	0.2832	0.2136	0.21
02085500	149	63	7585	0.2768	0.1733	0.1626	1.43
02333500	153	53	7807	0.3310	0.2449	0.1572	0.50
02228500	160	48	1978	0.3999	0.3889	0.3300	0.41

Note: HUC is the Hydrologic Unit Code of the basin. Area is the area of the basin, in square miles.
*denotes a D_i value that exceeds the critical value in Table 3.1.

3.4 Examples

As an example we use annual maximum streamflow data obtained from the U.S. Geological Survey (USGS). Data, measured in cubic feet per second, were obtained for a subset of the sites in the data set compiled by Wallis et al. (1991), and for this example we consider the sites in the southeastern U.S.A. (USGS Water Resources Region 3, as defined in Seaber, Kapinos, and Knapp, 1987). The 105 sites in Region 3 were assigned to six groups according to their drainage basin area. Record lengths and L-moment ratios for the 18 sites with smallest drainage basin areas are given in Table 3.2 and illustrated in Figure 3.2.

For the USGS Region 3 data, the D_i values are also given in Table 3.2. The critical value, 3, is exceeded by only one site: site 02267000, Catfish Creek near Lake Wales, Fla. This site has the lowest L-CV of any in the group, but it is not this fact alone that causes the high D_i value. As Figure 3.2 suggests, the discordancy arises because the combination of low L-CV and moderate L-skewness and L-kurtosis is discordant with the pattern of the other sites. In contrast, site 02111000, which has extremely high values of L-CV, L-skewness, and L-kurtosis, is not particularly discordant

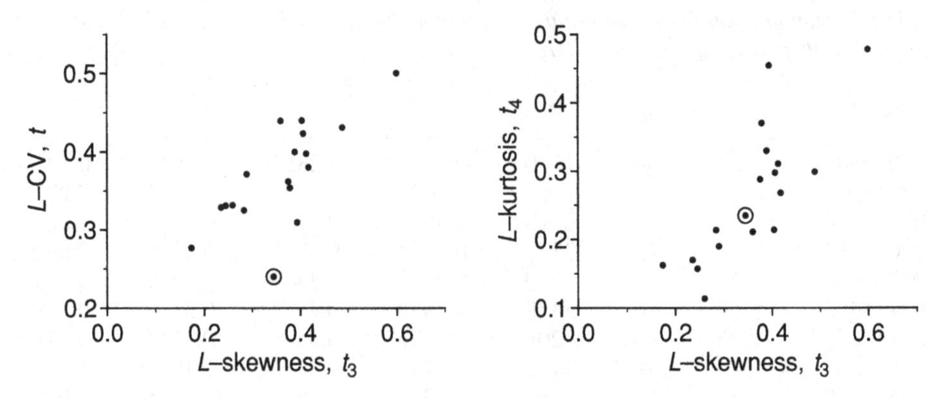

Fig. 3.2. L-moment ratios of the USGS Region 3 small-area sites. The circled point is for the site with $D_i > 3$.

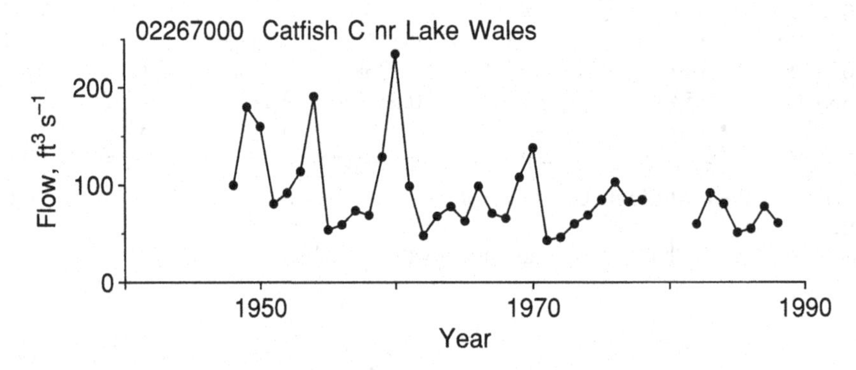

Fig. 3.3. Time-series plot of the data for site 02267000.

with the other sites. For this site, the deviation of the L-moment ratios from the group average, though large, is in a direction concordant with the corresponding deviations of the other sites, this direction being that of the major axis of the ellipsoid in Figure 3.1, and is downweighted by the $\mathbf{A}^{-1}$ matrix in the definition of D_i.

The data for the discordant site 02267000 are plotted in Figure 3.3 and show some evidence of a downward trend over time. Additionally, from Table 3.2 it is clear that the mean at this site is surprisingly low compared with other sites in the region with comparable drainage areas. There is, therefore, some reason to be suspicious of the data for this site. A final decision on whether to include this site in the analysis might require further investigation of whether the apparent trend can be attributed to random variation or whether it is a consequence of land-use changes or other man-induced effects that make the site an unreliable indicator of natural flow conditions. The missing values in 1979–81 should also be investigated.

Table 3.3. *Summary statistics for annual maximum streamflow data for sites in USGS Region 3 with large drainage basin area.*

HUC	Area	n	ℓ_1	t	t_3	t_4	D_i
02347500	1850	70	30611	0.3036	0.2103	0.1797	0.90
02083500	2183	87	15491	0.2617	0.2556	0.1771	0.33
02202500	2650	56	14600	0.3465	0.2818	0.2191	1.01
02478500	2690	51	25961	0.2710	0.2767	0.1465	1.54
02228000	2790	58	16426	0.3579	0.2635	0.1666	1.60
02135000	2790	47	13181	0.2435	0.1733	0.1618	1.62
02156500	2790	50	45374	0.2248	0.2754	0.2576	1.37
02349500	2900	84	30489	0.3100	0.2248	0.1708	0.52
02475000	3495	48	42463	0.2923	0.3059	0.2057	0.53
02365500	3499	61	36987	0.3017	0.4317	0.4164	3.38*
02488500	4993	65	39157	0.2491	0.2490	0.1926	0.40
02352500	5310	96	34739	0.2624	0.2059	0.1357	0.55
02489500	6573	50	48728	0.2570	0.2553	0.2409	0.66
02479000	6590	84	71568	0.2803	0.2655	0.1472	0.96
02320500	7880	57	19516	0.3173	0.2799	0.2506	0.65
02131000	8830	50	48636	0.2879	0.3476	0.2700	0.74
02226000	13600	64	69569	0.3149	0.2936	0.2231	0.23

Note: HUC is the Hydrologic Unit Code of the basin. Area is the area of the basin, in square miles.
*denotes a D_i value that exceeds the critical value in Table 3.1.

If they are missing for a reason related to the flow magnitude – for example, if a very large flood washed away the stream gage – then frequency analysis of the data must make allowance for this.

As a further example, we consider the sites in USGS Region 3 with the largest drainage areas. Record lengths and L-moment ratios for the 17 sites with the largest drainage basin areas are given in Table 3.3 and illustrated in Figure 3.4. The critical value, 3, is exceeded at only one site: site 02365500, Choctawhatchee River at Caryville, Fla. The site has moderate L-CV but very high L-skewness and L-kurtosis. A plot of the data shows that there was a very large flood in 1929. This outlying data value accounts for the high L-skewness and L-kurtosis for this site. The data value is plausible: The occurrence of a large flood in this year is confirmed by streamflow records at other nearby sites. For example, site 02349500, Flint River at Montezuma, Ga., 150 miles from Caryville, also had its largest flood on record in 1929. Plots of the data for these two sites are given in Figure 3.5. In this case there is no clear reason to doubt the validity of the data for the discordant site. This will often be the case in practice, because the variation of L-moment ratios between apparently similar sites can be quite large for many kinds of data.

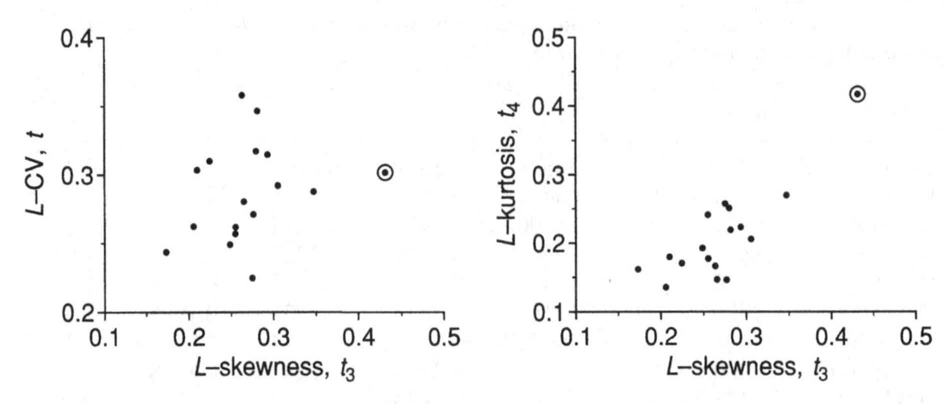

Fig. 3.4. L-moment ratios of the USGS Region 3 large-area sites. The circled point is for the site with $D_i > 3$.

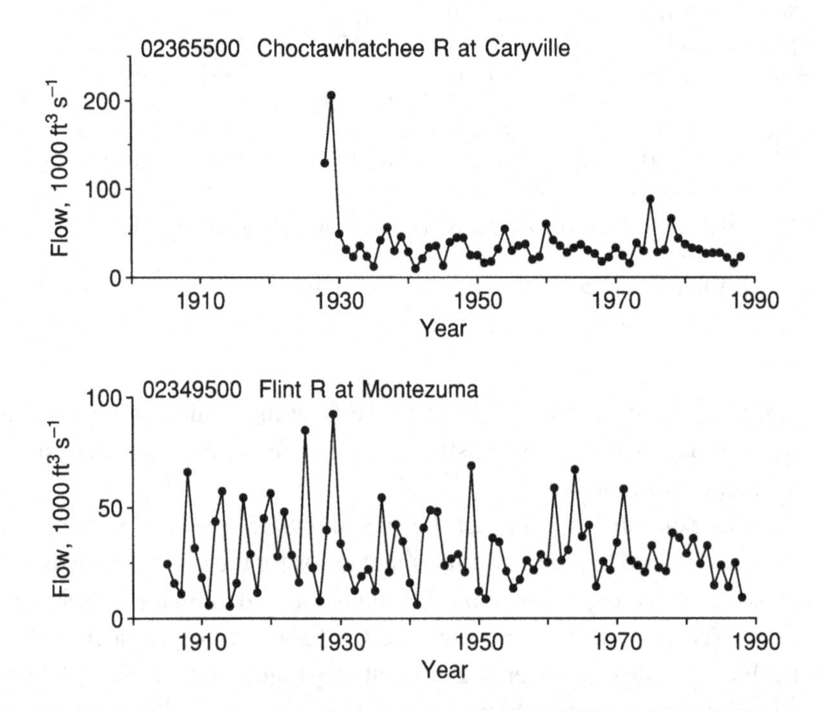

Fig. 3.5. Time-series plot of the data for sites 02365500 and 02349500.

A final example uses a set of annual precipitation totals obtained from the U.S. Historical Climatology Network (Karl et al., 1990). The data, in inches, are for the "North Cascades" region, one of 23 climatic divisions of the continental United States used by Plantico et al. (1990). Data are available for 19 sites. Record lengths and L-moment ratios are given in Table 3.4 and illustrated in Figure 3.6. Table 3.4 also shows the D_i values for the North Cascades data. None of the D_i values exceeds

Table 3.4. *Summary statistics for the North Cascades precipitation data.*

Site code	n	ℓ_1	t	t_3	t_4	t_5	D_i
350304	98	19.69	0.1209	0.0488	0.1433	−0.0004	0.60
351433	59	62.58	0.0915	0.0105	0.1569	0.0020	1.02
351862	90	40.85	0.1124	0.0614	0.1541	−0.0058	0.38
351897	61	46.05	0.1032	0.0417	0.1429	−0.0022	0.23
352997	65	45.02	0.0967	−0.0134	0.1568	0.0173	0.93
353445	86	31.04	0.1328	−0.0176	0.1206	0.0235	2.63
353770	78	80.14	0.1008	0.0943	0.1967	0.0856	2.12
356907	72	41.31	0.1143	0.0555	0.1210	0.0487	0.45
357169	67	30.59	0.1107	0.0478	0.1371	0.0316	0.11
357331	99	32.93	0.1179	0.0492	0.0900	0.0225	1.61
357354	49	17.56	0.1308	0.0940	0.1273	0.0352	2.08
358466	61	69.52	0.1119	−0.0429	0.0927	−0.0061	1.52
450945	69	47.65	0.1018	0.0435	0.1446	−0.0056	0.31
451233	73	102.50	0.1025	0.0182	0.1047	−0.0221	1.30
453284	70	52.41	0.1054	−0.0224	0.1664	0.0035	1.58
454764	66	79.70	0.1174	0.0124	0.1317	−0.0176	0.29
454769	59	44.64	0.1115	−0.0346	0.1032	0.0083	1.04
457773	74	58.66	0.1003	0.0446	0.1450	−0.0379	0.43
458773	82	39.02	0.1046	0.0128	0.1583	0.0443	0.38

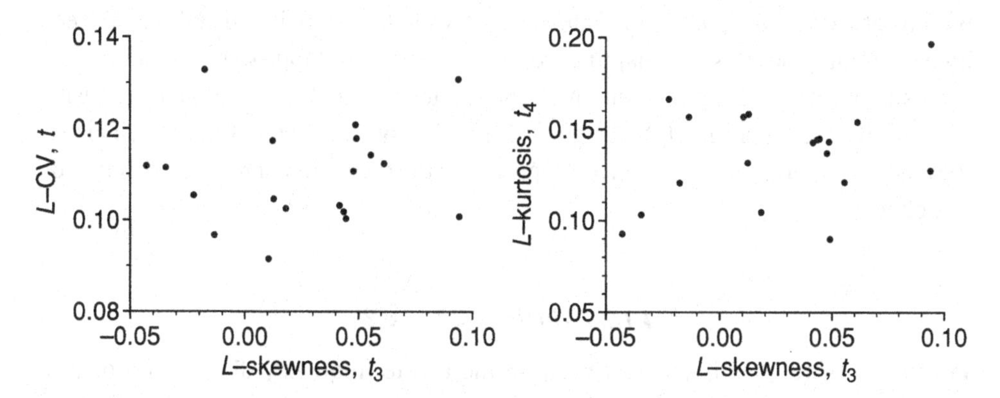

Fig. 3.6. *L*-moment ratios of the North Cascades sites.

the critical value, 3. The largest is 2.63 for site 353445, which has high *L*-CV and low *L*-skewness. It might be worthwhile to shift this site to another region if there are physical grounds for doing so, but there is no evidence of gross errors in the data.

4

Identification of homogeneous regions

4.1 Methods of forming regions

4.1.1 General considerations

Of all the stages in a regional frequency analysis involving many sites, the identification of homogeneous regions is usually the most difficult and requires the greatest amount of subjective judgement. The aim is to form groups of sites that approximately satisfy the homogeneity condition, that the sites' frequency distributions are identical apart from a site-specific scale factor. This is usually achieved by partitioning the sites into disjoint groups. An alternative approach is to define for each site of interest a region containing those sites whose data can advantageously be used in the estimation of the frequency distribution at the site of interest. This is the basis of the "region of influence" approach to the formation of regions, discussed in Section 8.1.

4.1.2 Which data to use?

Formation of regions is difficult because the at-site frequency distribution of the quantity of interest, Q, is not observed directly. The available data for region formation are quantities calculated from the at-site measurements of Q, which we call *at-site statistics*, and other site descriptors that we call *site characteristics*. In environmental applications the site characteristics would typically include the geographical location of the site, its elevation, and other physical properties associated with the site. Other site characteristics may be based on estimates rather than direct measurements, but are sufficiently accurate to be treated as though they were deterministic quantities. It is usually felt, for example, that mean annual precipitation can be reliably estimated from isohyetal maps prepared by state and government agencies. In principle, site characteristics are quantities that are known even before any data are measured at the site. However, it is reasonable to include among the site characteristics some quantities that are estimated from data measured at the site, provided that these measurements are not too highly correlated with the Q values themselves. Site characteristics of this kind might include the time of year at which the annual maximum event (flood or extreme precipitation) most frequently occurs. If quantile estimates are required at ungaged sites – a topic further discussed in Section 8.4 – it is advisable to use only site characteristics that can be reliably estimated when no at-site data are available.

We have made a distinction between at-site statistics and site characteristics, because we consider this distinction to be important. We strongly prefer to base the formation of regions on site characteristics and to use the at-site statistics only in subsequent testing of the homogeneity of a proposed set of regions. To understand why, suppose for the sake of concreteness that at-site sample L-CV values are used as the basis for forming regions, sites with similar values of L-CV being grouped together in the same region. L-CV is a critical variable for the formation of regions, because – as will be seen in Section 7.5.7 – it is the amount of dissimilarity between the L-CVs of the population distributions in a region that largely determines how much better regional frequency analysis will be than at-site analysis. There are three problems with defining regions based on the scatter of the sample L-CV values. First, little can be gained by using the regional estimate of L-CV from such a region, because the regional average L-CV will not be much different from any of the at-site sample L-CV values. Second, there is a tendency to group together all sites that have high outliers, and hence high L-CV, even though these outliers may be due to random fluctuations that happened to affect one site but not its neighbors. Third, we recommend that the homogeneity of the final regions be tested by a statistic that is calculated from the at-site statistics – the details are discussed in Sections 4.3–4.4. The integrity of this test is compromised if the same data are used both to form regions and to test them.

4.1.3 Grouping methods

Several authors have proposed methods for forming groups of similar sites for use in regional frequency analysis. The procedures can be roughly categorized as follows.

Geographical convenience

Regions are often chosen to be sets of contiguous sites, based on administrative areas (Natural Environment Research Council, 1975; Beable and McKerchar, 1982), or major physical groupings of sites (Matalas et al., 1975). Even though region boundaries may be adjusted after considering model fit (as in Schroeder and Massey, 1977), these approaches seem arbitrary and subjective and the resulting regions rarely give the impression of physical integrity.

Subjective partitioning

It is sometimes possible, particularly in small-scale studies, to define regions subjectively by inspection of the site characteristics. Schaefer (1990), analyzing annual maximum precipitation data for sites in Washington state, formed regions by grouping together sites with similar values of mean annual precipitation. Gingras, Adamowski, and Pilon (1994) formed regions for annual maximum streamflow data in Ontario and Quebec by grouping the sites according to the time of year at which the largest flood typically occurred.

Though methods such as these are subjective, the resulting regions can be objectively tested by the heterogeneity measure described in Section 4.3 below. A problem arises only if at-site statistics rather than site characteristics are used as the basis for the subjective partitioning. For example, Gingras and Adamowski (1993) formed regions for annual maximum streamflow data in New Brunswick, Canada, by grouping the sites according to whether a nonparametric estimate of the frequency distribution was unimodal, bimodal, or long-tailed. This procedure involves the at-site statistics to an extent that may affect the validity of the subsequent use of a heterogeneity measure to validate the regions.

Objective partitioning

In partitioning methods, regions are formed by assigning sites to one of two groups depending on whether a chosen site characteristic does or does not exceed some threshold value. The threshold is chosen to minimize a within-group heterogeneity criterion, such as a likelihood-ratio statistic (Wiltshire, 1985), within-group variation of the sample coefficient of variation $\hat{C}_v$ (Wiltshire, 1986a), or within-group variation of sample *L*-CV and *L*-skewness (Pearson, 1991a). The groups are then further subdivided in an iterative process until a final set of acceptably homogeneous regions is obtained. This approach is comparable to extrinsic hierarchical classification techniques of multivariate analysis, such as automatic interaction detection (Fielding, 1977), with which it shares the disadvantage that optimal choice of each successive dichotomous split of the set of sites need not yield an optimal final classification. Furthermore, the likelihood-ratio or within-group heterogeneity statistics are affected to an unknown degree when data at different sites are statistically dependent. There seems no reason to prefer this approach to the more standard cluster analysis methods, but it can be effective when used in conjunction with a subsequent assessment of whether the final regions are homogeneous, based, for example, on the heterogeneity measure defined in Section 4.3 below. Pearson (1991b) used this approach with Wiltshire's (1985) partitioning criterion and achieved a successful regionalization of streamflow data for small drainage basins in New Zealand.

Cluster analysis

Cluster analysis is a standard method of statistical multivariate analysis for dividing a data set into groups and has been successfully used to form regions for regional frequency analysis. A data vector is associated with each site, and sites are partitioned or aggregated into groups according to the similarity of their data vectors. The data vector can include at-site statistics, site characteristics, or some combination of the two; as noted above, we prefer to use only site characteristics.

De Coursey (1973) applied cluster analysis to site characteristics of streamflow gaging sites in Oklahoma to form groups of sites having similar flood response. Acreman and Sinclair (1986) analyzed annual maximum streamflow data for 168 gaging sites in Scotland and formed five regions, four of which they judged to be homogeneous. Burn (1989) used cluster analysis to derive regions for flood frequency analysis, though his clustering variables include at-site statistics. Guttman (1993) analyzed annual precipitation totals for 1119 sites in the U.S.A. and formed 104 regions, 101 of which were accepted as homogeneous; more details of this analysis are given in Section 9.1. Other examples of the use of cluster analysis in forming hydrological or climatological regions, albeit not for use in frequency analysis, have been given by Mosley (1981), Richman and Lamb (1985), Nathan and McMahon (1990), and Fovell and Fovell (1993). Farhan (1984) used cluster analysis to classify stream gaging sites in Jordan into regions on the basis of four principal components formed from a matrix of site characteristics.

We regard cluster analysis of site characteristics as the most practical method of forming regions from large data sets. It has several major variants and involves subjective decisions at several stages. Some suggestions for the use of cluster analysis in regional frequency analysis are given in the following subsection.

Other multivariate analysis methods

Other statistical multivariate analysis techniques have occasionally been used to form groups of similar sites. White (1975) used factor analysis of site characteristics to classify drainage basins in Pennsylvania. Burn (1988) used principal components analysis on series of annual maximum streamflow data and classified gaging sites according to which of a subjectively rotated set of the principal components a site's data record most closely resembled. This procedure is based on at-site statistics, but in a way that does not directly involve the shape of the at-site frequency distribution and therefore may not compromise the use of a statistical homogeneity test; however, for the same reason, it is not clear that the features of the at-site data that determine the grouping are useful in identifying sites that have similar frequency distributions.

4.1.4 Recommendations for cluster analysis

We regard cluster analysis of site characteristics as the most practical method of forming regions from large data sets. General descriptions of clustering methods are given by Gordon (1981) and Everitt (1993), and good discussions of the use of cluster analysis in environmental applications are given by Kalkstein, Tan, and Skindlov (1987) and Fovell and Fovell (1993). When the regions are intended for use in regional frequency analysis, some special considerations apply to cluster analysis.

Clusters are formed from groups of sites with similar site characteristics. Most clustering algorithms measure similarity by the reciprocal of Euclidean distance in a space of site characteristics. This distance measure is affected by the scale of measurement of the site characteristics, and in practice it is usual to rescale the site characteristics so that they all have the same amount of variability, as measured by their range or standard deviation across all of the sites in the data set. This rescaling effectively gives equal weight to each site characteristic in determining clusters; this may not be appropriate, because some site characteristics have a greater influence on the form of the frequency distribution and should be given greater weight in the clustering. It is difficult to choose appropriate weights. This is not a critical problem, however, because the validity of the final regions can be tested, by the method described in Section 4.3 below, without requiring that accurate weights be determined. Nonlinear transformation of the site characteristics may be appropriate too, to ensure that the influence of the site characteristic on the form of the frequency distribution is uniform across the range of values of the site characteristic.

There is no assumption that there are distinct clusters of sites that satisfy the homogeneity condition. More realistically, the form of the frequency distribution varies smoothly with the site characteristics, and the aim is to find groups of sites within which the site characteristics, and hence the at-site frequency distributions, vary so little that regional frequency analysis is preferable to at-site analysis and to regional frequency analysis based on any other set of regions. Thus there is no "correct" number of clusters; instead a balance must be sought between using regions that are too small or too large. Regions that contain few sites will achieve little improvement in the accuracy of quantile estimates over at-site analysis. Regions that cover a large part of the space of site characteristics may well fail to be homogeneous, causing bias in the quantile estimates at some of the sites.

These considerations have implications for the choice of clustering algorithm. Methods that tend to form clusters of roughly equal size should give good results. Examples of such algorithms are average-link clustering, which tends to form clusters with equal within-cluster variance of site characteristics, and Ward's method, which tends to form clusters containing equal numbers of sites. Methods that tend to form a small number of very large clusters, with a few small outlying clusters on the fringes of the space of site characteristics, are less likely to yield good regions for regional frequency analysis. Single-link, or nearest-neighbor, clustering is an example.

For regional frequency analysis with an index-flood procedure there is little advantage in using very large regions. Little gain in the accuracy of quantile estimates is obtained by using more than about 20 sites in a region – see Section 7.5.4. Thus there is no compelling reason to amalgamate large regions whose estimated regional frequency distributions are similar.

When cluster analysis is based on site characteristics, the at-site statistics are available for use as the basis of an independent test of the homogeneity of the final regions. Such tests are discussed in Sections 4.2–4.4. These tests directly address the aim of the cluster analysis and are a better guide to the choice of an appropriate number of clusters than the usual criteria of cluster analysis, such as the pseudo-t^2 criterion of Duda and Hart (1973) or the pseudo-F criterion of Calinski and Harabasz (1974).

The output from the cluster analysis need not, and usually should not, be final. Subjective adjustments can often be found to improve the physical coherence of the regions and to reduce the heterogeneity of the regions as measured by the heterogeneity measure H described in Section 4.3 below. Several kinds of adjustment of regions may be useful:

- move a site or a few sites from one region to another;
- delete a site or a few sites from the data set;
- subdivide the region;
- break up the region by reassigning its sites to other regions;
- merge the region with another or others;
- merge two or more regions and redefine groups; and
- obtain more data and redefine groups.

All of these adjustments were used in the example discussed in Section 9.1.

4.2 Tests of regional homogeneity

Once a set of physically plausible regions has been defined, it is desirable to assess whether the regions are meaningful. This involves testing whether a proposed region may be accepted as being homogeneous and whether two or more homogeneous regions are sufficiently similar that they should be combined into a single region.

The hypothesis of homogeneity is that the at-site frequency distributions are the same except for a site-specific scale factor. A test of this hypothesis is naturally based on whether the data at the N sites in the region are consistent with this relation between the at-site frequency distributions. The test is most conveniently constructed as a statistical significance test of the similarity of appropriately chosen statistics calculated from the at-site data. However, as we discuss in Section 4.4, such an interpretation of the test statistics should be used only with caution.

Tests of whether a region is homogeneous have been proposed by Dalrymple (1960), Acreman and Sinclair (1986), Wiltshire (1986a,b), Buishand (1989), Chowdhury, Stedinger, and Lu (1991), Lu and Stedinger (1992a), Hosking and Wallis (1993), and Fill and Stedinger (1995). Most of the tests involve a quantity θ that measures some aspect of the frequency distribution and is constant in a homogeneous region: θ may be the 10-year event scaled by the mean (Dalrymple, 1960; Lu and Stedinger, 1992a; Fill and Stedinger, 1995), the coefficient of variation (Wiltshire, 1986a), a combination of L-CV and L-skewness (Chowdhury et al., 1991), or the L-CV or some combination of L-CV, L-skewness, and L-kurtosis (Hosking and Wallis, 1993). Estimates of θ are calculated: $\hat{\theta}^{(i)}$ is an at-site estimate based on the data for site i, and $\hat{\theta}^{\mathrm{R}}$ is a regional estimate using data from all the sites in the region and assuming homogeneity. A test statistic S is then constructed that measures the difference between the at-site estimates and the regional estimate; one possible choice is

$$S = \sum_{i=1}^{N} (\hat{\theta}^{(i)} - \hat{\theta}^{\mathrm{R}})^2 . \tag{4.1}$$

This observed value of S is compared with the "null distribution" that S would have if the region were indeed homogeneous. The calculation of the null distribution usually involves an assumption about the form of the frequency distribution for the sites in the region. This distribution is assumed to be Gumbel by Dalrymple (1960) and Fill and Stedinger (1995), generalized extreme-value by Chowdhury et al. (1991) and Lu and Stedinger (1992a), and kappa by Hosking and Wallis (1993). Wiltshire (1986a) estimates the null distribution by a nonparametric jackknife procedure. If the observed value of S lies far in the tail of its null distribution, the hypothesis of homogeneity is rejected because it is deemed unlikely that so extreme a value of S could have arisen by chance from a homogeneous region.

Of the other approaches, Acreman and Sinclair (1986) and Buishand (1989) use likelihood-ratio tests that compare the fit of regional and at-site generalized extreme-value distributions fitted to the data by the method of maximum likelihood. Wiltshire's (1986b) test is based on the observation that if site i has data Q_{ij}, $j = 1, \ldots, n_i$, and a frequency distribution with cumulative distribution function $F_i(.)$, then the "G-statistics" $G_{ij} = F_i(Q_{ij})$, $j = 1, \ldots, n_i$, form a random sample from a uniform distribution on the interval (0, 1) and should take an average value of 0.5. When F_i is replaced by a fitted distribution obtained from regional analysis assuming homogeneity, uniformity of the G-statistics may be expected to hold approximately if the region really is homogeneous, but not otherwise. Wiltshire obtained a statistic that tests the similarity of the average deviations from 0.5 of each site's G-statistics calculated from the fitted regional distribution.

No general comparison of these tests has been made. All of the tests are statistically valid except that Dalrymple's calculation of the null distribution is invalid and has been corrected by Fill and Stedinger (1995). Some limited simulation experiments of Lu and Stedinger (1992a) and Fill and Stedinger (1995) found that tests based on L-moments outperform those of Wiltshire (1986a,b). We consider it inadvisable to use a test that makes too strict an assumption about the form of the frequency distribution. This is another aspect of the robustness discussed in Section 1.2. The tests of Hosking and Wallis (1993), which are based on L-moments and assume a kappa distribution – less restrictive than Gumbel or generalized extreme-value – for the underlying frequency distribution, can be generally recommended. These tests are described in Section 4.3.

4.3 A heterogeneity measure

4.3.1 Aim

The aim is to estimate the degree of heterogeneity in a group of sites and to assess whether the sites might reasonably be treated as a homogeneous region. Specifically, the heterogeneity measure compares the between-site variations in sample L-moments for the group of sites with what would be expected for a homogeneous region.

4.3.2 Heuristic description

In a homogeneous region all sites have the same population L-moment ratios. Their sample L-moment ratios will, however, be different, owing to sampling variability. Thus a natural question to ask is whether the between-site dispersion of the sample L-moment ratios for the group of sites under consideration is larger than would be expected of a homogeneous region. See Figure 4.1.

Let us consider how to measure the "between-site dispersion of sample L-moment ratios" and how to establish "what would be expected of a homogeneous region."

A visual assessment of the dispersion of the at-site L-moment ratios can be obtained by plotting them on graphs of L-skewness versus L-CV and L-skewness versus L-kurtosis. Reasonable numerical measures of dispersion based on these plots are the average distance from a site's plotted point on such a graph to the group average point. To allow for the greater variability of L-moment ratios in small samples, averages should be weighted proportionally to the sites' record lengths. An alternative and simple measure of the dispersion of the sample L-moment ratios is

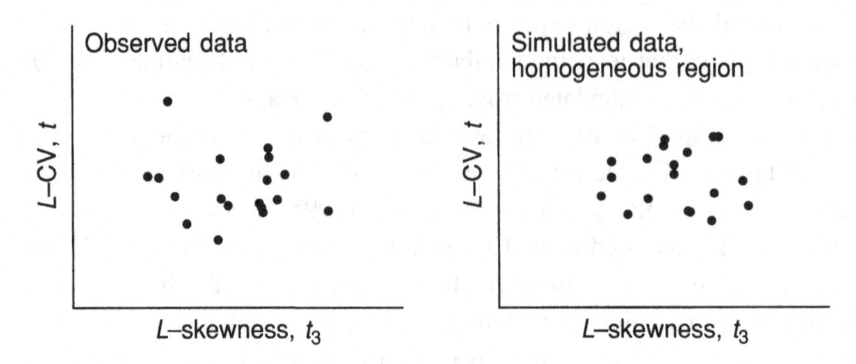

Fig. 4.1. Definition sketch for heterogeneity.

the standard deviation, again weighted proportionally to record length, of the at-site *L*-CVs. It is reasonable to concentrate on *L*-CV, as between-site variation in *L*-CV has a much larger effect than variation in *L*-skewness or *L*-kurtosis on the variance of the final estimates of all but the most extreme quantiles – see Section 7.5.7, in particular Table 7.11.

To establish "what would be expected" we use simulation. By repeated simulation of a homogeneous region with sites having record lengths the same as those of the observed data, we obtain the mean and standard deviation of the chosen dispersion measure. To compare the observed and simulated dispersions, an appropriate statistic is

$$\frac{(\text{observed dispersion}) - (\text{mean of simulations})}{(\text{standard deviation of simulations})} . \tag{4.2}$$

A large positive value of this statistic indicates that the observed *L*-moment ratios are more dispersed than is consistent with the hypothesis of homogeneity.

Finally, we must choose a distribution from which to generate the simulated data. If the observed sites do form a homogeneous region, this region's population *L*-moment ratios are likely to be close to the average of the sample *L*-moment ratios of the observed data. To avoid committing ourselves to a particular two- or three-parameter distribution, we use a four-parameter kappa distribution for the simulations. The kappa distribution, defined in Section A.10 of the appendix, includes as special cases the generalized logistic, generalized extreme-value, and generalized Pareto distributions. It is therefore capable of representing many of the distributions occurring in the environmental sciences. Its *L*-moments can be chosen to match the group average *L*-CV, *L*-skewness, and *L*-kurtosis of the observed data.

4.3.3 Formal definition

Suppose that the proposed region has N sites, with site i having record length n_i and sample L-moment ratios $t^{(i)}$, $t_3^{(i)}$, and $t_4^{(i)}$. Denote by t^{R}, t_3^{R}, and t_4^{R} the regional average L-CV, L-skewness, and L-kurtosis, weighted proportionally to the sites' record length; for example

$$t^{\mathrm{R}} = \sum_{i=1}^{N} n_i t^{(i)} \Big/ \sum_{i=1}^{N} n_i \,. \tag{4.3}$$

Calculate the weighted standard deviation of the at-site sample L-CVs,

$$V = \left\{ \sum_{i=1}^{N} n_i (t^{(i)} - t^{\mathrm{R}})^2 \Big/ \sum_{i=1}^{N} n_i \right\}^{1/2} . \tag{4.4}$$

Fit a kappa distribution to the regional average L-moment ratios 1, t^{R}, t_3^{R}, and t_4^{R} (see Section A.10 of the appendix).

Simulate a large number N_{sim} of realizations of a region with N sites, each having this kappa distribution as its frequency distribution. The simulated regions are homogeneous and have no cross-correlation or serial correlation; sites have the same record lengths as their real-world counterparts. For each simulated region, calculate V.

From the simulations determine the mean and standard deviation of the N_{sim} values of V. Call these μ_V and σ_V.

Calculate the heterogeneity measure

$$H = \frac{(V - \mu_V)}{\sigma_V} . \tag{4.5}$$

Declare the region to be heterogeneous if H is sufficiently large. We suggest that the region be regarded as "acceptably homogeneous" if $H < 1$, "possibly heterogeneous" if $1 \le H < 2$, and "definitely heterogeneous" if $H \ge 2$.

4.3.4 Performance

The performance of H as a heterogeneity measure was assessed in a series of Monte Carlo simulation experiments. For each of a number of artificial regions, 100 replications were made of data from the region, and the accuracy of quantile estimates and values of the heterogeneity measure H were calculated. N_{sim}, the number of regions simulated in the computation of H, was 500. Simulation results are given in Table 4.1. Regions were specified by the number of sites in the region,

Table 4.1. *Simulation results for the heterogeneity measure H.*

Region type	L-CV		No. of sites	RMSE of quantiles				Ave. of het. measures	
	Ave.	Range		.01	.1	.99	.999	H	H^*
Hom $n=30$	.20	0	6	9.4	7.1	8.7	11.7	0.11	1.03
	.20	0	11	8.2	6.8	8.1	10.5	0.10	1.02
	.20	0	21	7.8	6.8	7.9	10.0	0.06	1.01
	.30	0	21	14.0	11.2	12.8	16.1	0.02	1.00
	.15	0	21	5.9	5.0	5.4	6.5	0.00	1.00
Het 30% $n=30$	.20	.06	6	12.9	8.6	12.5	18.1	0.91	1.30
	.20	.06	11	12.3	8.7	11.0	15.8	1.08	1.25
	.20	.06	21	11.5	8.5	10.8	15.3	1.19	1.19
	.30	.09	2	19.8	14.1	16.4	23.2	1.07	1.19
	.15	.045	21	7.6	6.0	8.6	12.1	1.37	1.22
Het 50% $n=30$	.20	.10	6	18.2	11.3	16.9	25.1	2.09	1.69
	.20	.10	11	17.4	11.4	15.0	22.2	2.51	1.58
	.20	.10	21	16.3	10.9	14.6	21.5	2.96	1.48
	.30	.15	21	28.1	18.5	21.3	32.2	2.49	1.44
	.15	.075	21	9.9	7.5	12.3	18.0	3.41	1.54
Hom $n=60$	.20	0	6	6.8	5.1	6.8	9.4	–0.02	0.99
	.20	0	11	5.8	4.9	5.9	7.6	0.28	1.06
	.20	0	21	5.4	4.8	5.3	6.3	0.07	1.01
Het 30% $n=60$	.20	.06	6	12.0	7.7	10.7	16.3	1.55	1.51
	.20	.06	11	10.8	7.3	9.5	13.9	2.16	1.49
	.20	.06	21	10.3	7.1	8.9	12.9	2.41	1.39

the record lengths at each site, and the frequency distribution at each site. Frequency distributions were generalized extreme-value at each site and were specified by their L-moment ratios τ and τ_3; the at-site mean was, without loss of generality, set to 1 at each site. Three types of region were used in the simulations: homogeneous; heterogeneous, with L-CV and L-skewness varying linearly from site 1 through site N; and "bimodal," with half the sites having one distribution and half another. These regions test the ability of H to detect heterogeneity both when the frequency distributions vary smoothly from site to site and when there is a sharp difference between the frequency distributions at two subsets of sites. The base region for the simulations has $N=21$ and $n_i=30$ at each site and regional average values of 0.2 for both τ and τ_3. Variations on this region include changing N to 6 or 11, changing each n_i to 60, and changing the regional average τ to 0.1 or 0.3 with appropriate changes in τ_3. Both homogeneous and heterogeneous variants of these regions were simulated.

Table 4.1. *(continued)*

Region type		*L*-CV Ave.	*L*-CV Range	No. of sites	RMSE of quantiles .01	.1	.99	.999	Ave. of het. measures *H*	H^*
Het 50% $n = 60$		.20	.10	6	18.1	11.0	15.5	24.0	3.53	2.16
		.20	.10	11	16.3	10.3	13.8	20.9	4.51	2.03
		.20	.10	21	15.8	10.1	13.2	19.9	5.45	1.87
Het 30%	(a)	.20	.06	21	12.8	9.9	11.7	15.8	1.37	1.22
	(b)	.20	.06	21	12.0	8.7	11.6	16.4	0.80	1.13
	(c)	.20	.06	21	12.1	9.1	11.1	15.1	1.63	1.27
	(d)	.20	.06	21	12.3	9.5	11.0	14.8	0.79	1.13
Hom $n = 30$		.20	0	2	13.1	7.5	12.8	21.6	0.00	1.00
		.20	0	4	10.7	7.3	10.6	15.7	–0.01	0.99
		.20	0	10	8.4	6.7	8.6	11.9	0.11	1.02
		.20	0	20	7.7	6.9	7.8	9.6	0.13	1.02
Bimodal 20% $n = 30$		.20	.04	2	16.8	9.5	15.0	25.4	0.62	1.46
		.20	.04	4	14.3	9.1	13.3	20.5	0.71	1.30
		.20	.04	10	12.6	8.7	11.8	17.2	1.00	1.24
		.20	.04	20	12.0	8.8	11.3	15.8	1.56	1.26
Bimodal 30% $n = 30$		.20	.06	2	20.3	11.5	17.5	29.6	1.27	1.97
		.20	.06	4	17.9	11.0	16.1	25.0	1.48	1.63
		.20	.06	10	16.5	10.8	14.8	21.9	2.03	1.50
		.20	.06	20	16.0	10.8	14.4	20.9	3.02	1.51
Bimodal 50% $n = 30$		.20	.10	2	28.8	16.4	24.0	39.9	2.83	3.16
		.20	.10	4	26.3	16.0	22.9	35.7	3.30	2.42
		.20	.10	10	25.4	15.8	21.6	32.6	4.59	2.13
		.20	.10	20	25.2	15.9	21.6	32.2	6.57	2.10

Note: "Region type" is Homogeneous, Heterogeneous (*L*-CV τ and *L*-skewness τ_3 increase linearly from site 1 to site N), or Bimodal (half the sites have high τ and τ_3, the other half have low τ and τ_3). All regions have generalized extreme-value frequency distributions. "Het 30%" means that (range of τ) $\div$ (average τ) is 0.3. Columns 2 and 3 are the average and the range of τ for the region. Here, τ_3 is equal to τ at all sites, except when average τ is 0.15; in this case average τ_3 is 0.1 and the range of τ_3 is 0.09 for the Het 30% region and 0.15 for Het 50%. "RMSE of quantiles" is the relative RMSE of estimated quantiles, expressed as a percentage; tabulated values are calculated from 100 simulations. Sample size: $n_i = 30$ or $n_i = 60$ at each site, where indicated; Region (a) has $n_i = 50, 48, \ldots, 10$ at sites $i = 1, 2, \ldots, 21$; Region (b) has $n_i = 10, 12, \ldots, 50$; Region (c) has $n_i = 50, 46, \ldots, 14, 10, 14, \ldots, 46, 50$; and Region (d) has $n_i = 10, 14, \ldots, 46, 50, 46, \ldots, 14, 10$.

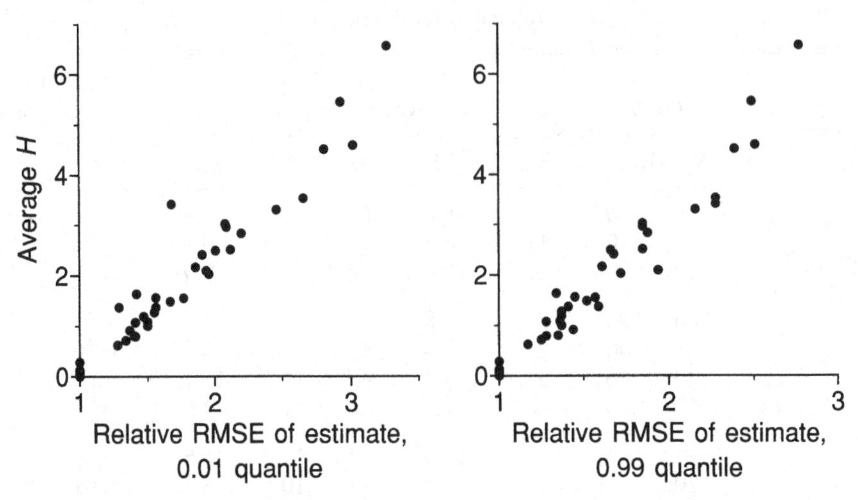

Fig. 4.2. Average H value and relative RMSE of quantile estimates for the simulated regions of Table 4.1.

Quantile estimates were obtained by regional frequency analysis, fitting a generalized extreme-value distribution using the regional L-moment algorithm described in Section 6.2. The relative RMSE of the quantile estimate $\hat{Q}_i(F)$ was calculated for each site. The "RMSE of quantiles" in Table 4.1 is the average over all sites in the region of this relative RMSE, expressed as a percentage. The "Ave. of het. measures" columns give the average, over the 100 replications of each region, of the heterogeneity statistic H and of a variant H^* described below in Subsection 4.3.7.

For purposes of estimating extreme quantiles, we consider the true measure of heterogeneity to be the amount by which the error in the quantile estimates is greater for the observed region than for a homogeneous region with the same values for N, the n_i, and the regional average L-moment ratios. This error cannot be calculated for observed data because the underlying frequency distribution is unknown but can be found for simulated data and can be calculated from Table 4.1. Figure 4.2 summarizes the relationship between the average H value for a simulated region and the RMSE of quantile estimates for that region relative to a homogeneous region. In general, the relationship is fairly well defined, showing that H is indeed a reasonable proxy for the likely error in quantile estimates. The $H = 1$ level is reached when the RMSE is 20–40% higher than for a homogeneous region; $H = 2$ is reached when the RMSE is 40–80% higher than for a homogeneous region. The main doubt concerning the H measure is excessive dependence on the number of sites in the region, particularly when the focus is on estimating quantiles that are not really extreme. In the "Bimodal" regions, for example, the RMSE relative to a homogeneous region for the $F = 0.1$ quantile varies very little as the number N of sites in the region varies, but the average H value decreases steadily as N

decreases. This means that H is better at indicating heterogeneity in large regions but has a tendency to give false indications of homogeneity for small regions. This effect is less marked at more extreme quantiles.

4.3.5 Notes

The assessment of heterogeneity by comparison of L-moments of observed data with those of data simulated from a homogeneous region has sometimes been made informally, using only one or two simulated regions (Hosking, 1990, Fig. 7; Pearson, 1991a, 1993; Pilon, Adamowski, and Alila, 1991; Pilon and Adamowski, 1992; Wallis, 1993). The use of H with a large number of simulations is a less subjective variant of this approach.

The value of N_{sim} should be chosen to achieve reliable estimates μ_V and σ_V. From simulations we judge that a value of $N_{\text{sim}} = 500$ should usually be adequate. Larger values may be need to resolve H values very close to 1 or 2.

The use of a kappa distribution in the simulations is, as noted above, intended to avoid too early a commitment to a particular distribution as the parent of the observed data. This contrasts with homogeneity tests proposed by Acreman and Sinclair (1986) and Chowdhury et al. (1991), which involve fitting generalized extreme-value distributions to the data. Using these tests, when the homogeneity hypothesis is rejected, it remains doubtful whether the region is heterogeneous or whether it is homogeneous but has some other frequency distribution.

It may not be possible to fit a kappa distribution to the group average L-moments. This occurs if t_4^{R} is too large relative to t_3^{R}. In such cases we recommend that the generalized logistic distribution, a special case of the kappa distribution with parameter h equal to -1, be used for the simulated region.

4.3.6 Alternative measures of dispersion

It is possible to construct heterogeneity measures in which V in Eq. (4.4) is replaced by other measures of between-site dispersion of sample L-moments. We considered a measure based on L-CV and L-skewness

$$V_2 = \sum_{i=1}^{N} n_i \{(t^{(i)} - t^{\text{R}})^2 + (t_3^{(i)} - t_3^{\text{R}})^2\}^{1/2} \Big/ \sum_{i=1}^{N} n_i \tag{4.6}$$

and a measure based on L-skewness and L-kurtosis

$$V_3 = \sum_{i=1}^{N} n_i \{(t_3^{(i)} - t_3^{\text{R}})^2 + (t_4^{(i)} - t_4^{\text{R}})^2\}^{1/2} \Big/ \sum_{i=1}^{N} n_i \,. \tag{4.7}$$

V_2 and V_3 are the weighted average distance from the site to the group weighted mean on graphs of t versus t_3 and of t_3 versus t_4, respectively. For both real-world data and artificial simulated regions, H statistics based on V_2 and V_3 lack power to discriminate between homogeneous and heterogeneous regions: They rarely yield H values larger than 2 even for grossly heterogeneous regions. The H statistic based on V has much better discriminatory power. Similar results have been reported by Lu (1991).

The measure V is, of course, insensitive to heterogeneity that takes the form of sites having equal L-CV but different L-skewness, but this form of heterogeneity has little effect on the accuracy of quantile estimates except very far into the extreme tails of the distribution and is in any case rare in practice, because, for most kinds of data, sites with high L-skewness tend to have high L-CV too. Thus we judge that V is clearly superior to V_2 and V_3 for the between-site comparisons of sample L-moment ratios needed by the index-flood procedure.

Some regionalization procedures seek to define regions in which L-skewness and L-kurtosis are constant while L-CV may vary. These procedures include the "regional shape estimation" and "hierarchical regions" approaches discussed in Section 8.1. The measure based on V_3 should be an appropriate tool for assessing the heterogeneity of proposed regions when using these procedures.

4.3.7 An alternative heterogeneity measure

The heterogeneity measure H defined in Eq. (4.5) is constructed like a significance test of whether all the sites in the region have, after scaling by division by the mean, identical frequency distributions. A region will be accepted as homogeneous if the data are consistent, at a suitable level of significance, with the hypothesis that the at-site population L-CVs are identical. Such a region will be said to be *statistically homogeneous*. In contrast, however, what is required for regional frequency analysis is that the region be *operationally homogeneous*, that is, that the at-site population L-CVs, although not necessarily identical, should be sufficiently close to each other that regional analysis is more accurate when applied to the region as a whole than when applied separately to two or more subregions.

The difference between statistical homogeneity and operational homogeneity can be important. For example, when a region has a small number of sites there are only a small number of at-site sample L-CVs available for use in a significance test, and the differences between them can be large and yet not be statistically significant. Thus the region is likely to be accepted as statistically homogeneous even when it is not operationally homogeneous. Conversely, when a proposed region has many

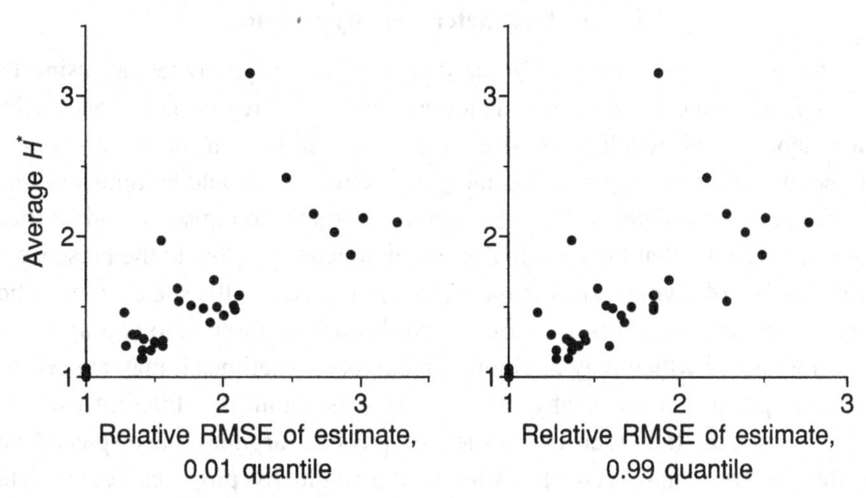

Fig. 4.3. Average H^* value and relative RMSE of quantile estimates for the simulated regions of Table 4.1.

sites then although the spread of the population L-CVs may be small enough for the region to be operationally homogeneous, it is unlikely that the sample L-CVs would be close enough for the region to be accepted as statistically homogeneous. To overcome this difficulty, a test must be constructed that is truly sensitive to operational heterogeneity rather than statistical heterogeneity. One possible approach is to test the hypothesis that the variation in the at-site population L-CVs is less than some fixed threshold at which regional and at-site estimation would be equally accurate and to apply this test at a significance level of 50%. A suitable heterogeneity measure can be devised; follow the same procedure as given in Subsection 4.3.3 but in place of H in Eq. (4.5) define

$$H^* = V/\mu_V. \tag{4.8}$$

H^* is the ratio of the standard deviation of the observed at-site L-CVs to the average value of the same standard deviation for the simulated kappa regions.

Although H^* has a theoretical justification as given above, its practical performance is disappointing. It was tested on simulated data in the same way as H, and results for it are included in Table 4.1. Figure 4.3 compares the average H^* value for a simulated region and the RMSE of quantile estimates for that region relative to a homogeneous region. Figures 4.2 and 4.3 enable comparison of H and H^*. The relationship between the heterogeneity measure and the accuracy of quantiles is better defined for H than for H^*, so it is preferable to use H.

4.4 Use of the heterogeneity measure

The heterogeneity measure may be used to assess a proposed region, using the value of H to judge the degree of heterogeneity. If the region is not acceptably homogeneous, some redefinition of the region should be considered. The region could be divided into two or more subregions, some sites could be removed from the region, or a completely different assignment of sites to regions could be tried. However, it may be that the appearance of heterogeneity is due to the presence of a small number of atypical sites in the region. In such cases the site characteristics for these sites should be carefully examined. It may be possible to reassign these sites to regions of which they are more typical, but sometimes it may appear that there is no physical reason why the atypical sites should be different from the rest of the region. It is then best to let the physical argument take precedence over the statistical and retain the sites in the originally proposed region. This approach makes possible the identification of homogeneous regions even when the homogeneity is masked by sampling variation in the data. For example, a certain combination of extreme meteorological conditions and consequent extreme events may be capable of occurring at any of a number of sites in a region but have actually occurred at only a few of these sites during the period of measurement. The greatest potential benefits of regionalization can be attained if in such a situation physical knowledge enables the entire set of sites to be identified as a homogeneous region. In this case, the at-site data are misleading, being unduly influenced by the presence or absence of an unusual environmental event, and the regional average frequency distribution will give the best estimates of the future risk of extreme events.

The H statistic is constructed like a significance test of the hypothesis that the region is homogeneous. However, we do not recommend that it be used in this way. Significance levels obtained from such a test would be accurate only under special assumptions: that the data are independent both serially and between sites, and that the true regional distribution is kappa. We need to define a heterogeneity measure for regions that may not satisfy these assumptions, so we prefer not to use H as a significance test. It would be possible to generate simulated data that are correlated, but this would require much more computing time. A significance test is of doubtful utility anyway, because even a moderately heterogeneous region can provide quantile estimates of sufficient accuracy for practical purposes. Thus a test of exact homogeneity is of little interest.

The criteria $H = 1$ and $H = 2$ are somewhat arbitrary, but we believe them to be useful guidelines. If H were used as a significance test, assuming the V statistic to be Normally distributed, then the criterion for rejection of the hypothesis of homogeneity at significance level 10% would be $H = 1.28$. A criterion of $H = 1$

may seem very strict in comparison, but, as noted above, we do not seek to use H in a significance test. From the simulation results, a region sufficiently heterogeneous that quantile estimates for it are 20–40% less accurate than for a homogeneous region will on average yield $H \approx 1$. We regard this amount of heterogeneity as being on the borderline of whether a worthwhile increase in the accuracy of quantile estimates could be achieved by redefining the region. For such a region, it is still likely that regional estimation will yield much more accurate quantile estimates than at-site estimation, but it is possible that subdividing the region or removing a few sites from it may reduce its heterogeneity. We therefore regard $H = 1$ as the limit at which seeking to redefine the region may be advantageous. Similarly, we regard $H = 2$ as a point at which redefining the region, if the available site characteristics permit it, is very likely to be beneficial.

Negative values of H are sometimes observed. These indicate that there is less dispersion among the at-site sample L-CV values than would be expected of a homogeneous region with independent at-site frequency distributions. The most likely cause is positive correlation between the data values at different sites. If many large negative values, $H < -2$ say, are obtained during a regional analysis, this may be an indication that there is a large amount of cross-correlation between the sites' frequency distributions or that there is some excessive regularity in the data that causes the sample L-CVs to be unusually close together. Further examination of the data would then be warranted.

We emphasize again that the validity of H as a heterogeneity measure is compromised if the selection of regions is based on sample L-moments, for then the same data are being used both to identify regions and to test their homogeneity. One could, for example, define a region to consist of all sites with a sample L-CV within a certain small range. Such a region might yield a small value of H, but this would reflect only the pattern of noise, or sampling variability, in the data and have no physical significance. Valid use of H requires that assignment of sites to regions be based on external site characteristics such as the physical characteristics or geographical location of the sites.

4.5 Example

We calculate the heterogeneity statistic for the North Cascades data in Table 3.4. The V measure calculated from the observed data is 0.01044. The group average L-moments are

$$t^{\mathrm{R}} = 0.1103, \qquad t_3^{\mathrm{R}} = 0.0279, \qquad t_4^{\mathrm{R}} = 0.1366, \tag{4.9}$$

and the parameters of the fitted kappa distribution are

$$\xi = 0.9542, \qquad \alpha = 0.1533, \qquad k = 0.1236, \qquad h = -0.2955. \tag{4.10}$$

Simulations were made of this kappa region. The V measures for 500 simulated regions had an average of 0.00948 and a standard deviation of 0.00156. The calculated heterogeneity measure H is thus $(0.01044 - 0.00948)/0.00156 = 0.62$, and the region is acceptably homogeneous.

5
Choice of a frequency distribution

5.1 Choosing a distribution for regional frequency analysis

5.1.1 General framework

In regional frequency analysis a single frequency distribution is fitted to data from several sites. In general, the region will be slightly heterogeneous, and there will be no single "true" distribution that applies to each site. The aim is therefore not to identify a "true" distribution but to find a distribution that will yield accurate quantile estimates for each site.

The chosen distribution need not be the distribution that gives the closest approximation to the observed data. Even when a distribution can be found that gives a close fit to the observed data, there is no guarantee that future values will match those of the past, particularly when the data arise from a physical process that can give rise to occasional outlying values far removed from the bulk of the data. As noted in Section 1.2, it is preferable to use a robust approach based on a distribution that will yield reasonably accurate quantile estimates even when the true at-site frequency distributions deviate from the fitted regional frequency distribution.

There may be a particular range of return periods for which quantile estimates are required. In analyses of extreme events such as floods or droughts, quantile estimates in one tail of the distribution will be of particular interest. In other examples, quantiles far into the tails of the distribution may be of little interest. These considerations may affect the choice of regional frequency distribution. If only quantiles in the upper tail are of interest, for example, then it need not matter if a distribution that can take negative values is fitted to data that can only be positive.

5.1.2 Selection of candidate distributions

There are many families of distributions that might be candidates for being fitted to a regional data set. Their suitability as candidates can be evaluated by considering their ability to reproduce features of the data that are of particular importance in modeling. The following properties of a distribution may be of importance in any given application.

Upper bound of the distribution

Many physical quantities can be thought of as having an upper bound. The bound may not be known exactly, but some numerical values are so unlikely as to be physically impossible. For example, windspeeds of 1,000 mi hr^{-1} (600 km hr^{-1}) or instantaneous rainfall rates of 20 in hr^{-1} (500 mm hr^{-1}) would be considered physically impossible. For this reason it is sometimes argued (e.g., Boughton, 1980; Laursen, 1983) that only bounded distributions should be used. We maintain that this argument is misguided. If the aim of an analysis is to estimate quantiles of return periods up to 100 years, that the estimated quantile at return period 100,000 years is "physically impossible" is of no relevance and should not be any cause for concern. Indeed, imposing the requirement that the distribution have a physically realistic upper bound may compromise the accuracy of quantile estimates at the return periods that are of real interest. When an unbounded distribution is used, the assumption, usually implicit, may be that the upper bound of the distribution cannot be estimated with sufficient accuracy to be worth the effort, or that over the range of return periods of interest in the particular study the true distribution function is likely to be better approximated by an unbounded distribution than by any easily parametrizable bounded distribution.

Of course, when it appears that the true distribution has an upper bound that is closely approached by the observed data, then it is advisable to fit a distribution that is capable of modeling bounded data. For example, the generalized extreme-value distribution has an upper bound when its shape parameter k is greater than zero. When this distribution is fitted to data, a tendency for the data to lie close to an upper bound will be reflected in an estimated k value greater than zero.

Upper tail of the distribution

In many applications, estimation of the upper tail of the frequency distribution is of particular interest, yet the amount of data is not sufficient to determine the shape of the upper tail with any accuracy. The tail weight, the behavior of the probability density function $f(x)$ as x increases, is important because it determines the rate at which quantiles increase as the return period is extrapolated beyond the range of the data. Tail weights of some common distributions are given in Table 5.1. When there is no reason to assume that only one kind of tail weight is appropriate, it is advisable to use a set of candidate distributions that cover a range of different tail weights. The goodness-of-fit statistic described in Section 5.2 provides a means of

Table 5.1. *Upper-tail weights of some common distributions.*

Form of $f(x)$ for large x	Distributions
x^{-A}	Generalized extreme-value, generalized Pareto, and generalized logistic distributions with parameter $k < 0$.
$x^{-A \log x}$	Lognormal distribution with positive skewness.
$\exp(-x^A)$, $0 < A < 1$	Weibull distribution with parameter $\lambda < 1$.
$x^A e^{-Bx}$	Pearson type III distribution with positive skewness.
$\exp(-x)$	Exponential, Gumbel.
$\exp(-x^A)$, $A > 1$	Weibull distribution with parameter $\lambda > 1$.
Finite upper bound	Generalized extreme-value, generalized Pareto, and generalized logistic distributions with parameter $k > 0$; lognormal and Pearson type III distributions with negative skewness.

Note: Tail weights are ordered from heaviest to lightest. A and B denote arbitrary positive constants.

deciding which distributions, and hence which tail weights, are consistent with a set of homogeneous regional data.

Shape of the body of the distribution

Most distributions used in statistics have a probability density function with a single peak from which the density declines smoothly in both directions. In some analyses there may be reason to consider other distributions. For example, annual maximum streamflow data may contain some values arising from floods caused by snowmelt and some from rainstorm events. In such cases, when the data are observations of a phenomenon that may arise from qualitatively different causes, a mixture of two distributions may be entertained; the regional frequency distribution would be

$$F(x) = pG_1(x) + (1 - p)G_2(x), \tag{5.1}$$

where $G_1(x)$ and $G_2(x)$ are the cumulative distribution functions of data arising from the two distinct causes and p is the proportion of the observations that arise from the first cause. There is no theoretical reason why such a distribution cannot be used in a regional L-moment analysis. There may bc practical difficulties, however, insofar as expressions for probability weighted moments and L-moments of mixed distributions tend to be complicated. See, for example, the results for the "two-component extreme-value distribution" given by Beran, Hosking, and Arnell (1986).

Lower tail of the distribution

Similar considerations apply to the lower tail of the distribution as to the upper tail; in many cases it is advisable to consider a range distributions with different tail weights. If interest centers on the upper tail of the distribution, however, the form of the lower tail is irrelevant. For example, an exponential distribution may give a good approximation to the upper-tail quantiles of annual maximum streamflow data (Damazio and Kelman, 1986), even though the shape of the lower tail of the distribution bears little resemblance to that of the data.

Lower bound of the distribution

Similar considerations apply to the lower bound of the distribution as to the upper bound; even if a lower bound exists, it may not be efficient to try to estimate it explicitly. Unlike the upper bound, however, the lower bound may often be known; usually it will be known to be zero. If quantiles of interest are close to zero, it may be worthwhile to require the lower bound of the regional frequency distribution to be zero. Several distributions, such as the Wakeby, generalized Pareto, and Pearson type III, retain a convenient form when this requirement is imposed. In some cases, knowledge that the lower bound is zero is not useful, and better results will be obtained by fitting a distribution that has a lower bound greater than zero or even a distribution that has no lower bound. For example, annual precipitation totals are bounded below by zero, but in most temperate parts of the world values close to zero are so unlikely that a realistic distribution of annual precipitation totals will have a lower bound considerably greater than zero.

Exact zero values

Some data, such as precipitation totals, may contain a number of zero values. If estimates of quantiles in the lower tail of the distribution are important, a distribution that allows for a nonzero proportion of zero values should be used. Suitable distributions can be obtained from existing standard distributions. One might, for example, use a generalized extreme-value distribution adjusted so that the fraction of the distribution that is negative is replaced by an atom of probability at zero. Unfortunately, the L-moments of such adjusted distributions tend to be difficult to work with. Wang (1990a) and Hosking (1995) have given some L-moment calculations for the related problem of inference from censored distributions.

Alternatively, a mixed distribution may be used, with the assumed regional frequency distribution having the form

$$F(x) = \begin{cases} 0, & x < 0, \\ p + (1 - p)G(x), & x \geq 0. \end{cases} \tag{5.2}$$

Here p is the probability of a zero value and $G(x)$ is the cumulative distribution function of the nonzero values; it may or may not be constrained to have a lower bound of exactly zero. The p parameter can be estimated by the proportion of zero values in the data for the region, and the distribution $G(x)$ can be fitted using the regional L-moments of the nonzero data values. This approach was used by Guttman, Hosking, and Wallis (1993) and is described in Section 9.1.

Sometimes there are theoretical reasons why a particular family of distributions is appropriate for a given type of data. Since the work of Gumbel (1958) it has been argued that data on extreme events such as annual maximum streamflows or precipitation may be well fitted by extreme-value distributions. For annual maximum streamflows, for example, the extreme-value approximation is valid when in each year there are a large number of storm events whose peak streamflows are independent and identically distributed. In practice, the assumptions underlying the extreme-value approximation may not be satisfied. For annual maximum streamflow data, the number of storm events in a year is rarely large enough to justify the extreme-value approximation, and the storm event magnitudes, rather than being identically distributed, tend to vary with the seasons of the year. Though an extreme-value distribution may be a candidate for describing the data, it should not be used without comparing its goodness of fit with that of other distributions.

Some thought should be given to the number of unknown parameters in the candidate distributions. Distributions with only two parameters yield accurate quantile estimates when the true distribution resembles the fitted distribution, but estimates of tail quantiles can be severely biased if the shape of the tail of the true frequency distribution is not well approximated by the fitted distribution – see Section 7.5.8. The use of a distribution with more parameters, when these can be estimated accurately, yields less biased estimates of quantiles in the tails of the distribution. One of the advantages of regional frequency analysis is that distributions with three or more parameters can be estimated more reliably than would be possible using only a single site's data. The use of a Wakeby distribution, with five parameters, when the sample size at each site is only 20, for example, is perfectly reasonable. For most applications of regional frequency analysis we feel that distributions with three to five parameters are appropriate.

5.1.3 Final choice of a regional frequency distribution

Assessment of the merits of different candidate distributions for a particular application will largely be based on how well the distributions fit the available data. In general, one would not want to use a distribution that is inconsistent with the data. However, this is not to say that one should invariably choose the distribution that gives the best fit to the data. The aim of regional frequency analysis is not to fit a particular data set but to obtain quantile estimates of the distribution from which future data values will arise. When several distributions fit the data adequately, any of them is a reasonable choice for use in the final analysis, and the best choice from among them will be the distribution that is most robust, that is, most capable of giving good quantile estimates even though future data values may come from a distribution somewhat different from the fitted distribution.

Several methods are available for testing the goodness of fit of a distribution to data from a single sample. These include quantile–quantile plots, chi-squared, Kolmogorov–Smirnov, and other general goodness-of-fit tests and tests based on moment or *L*-moment statistics. Some of these methods can be adapted for use in the regional framework. The fit of a postulated regional frequency distribution to each site's data can be assessed by goodness-of-fit statistics calculated at each site, and the resulting statistics then combined into a regional goodness-of-fit statistic. This is a reasonable approach and has been used by Chowdhury et al. (1991).

Using *L*-moments, it is natural to base test statistics on at-site and regional *L*-moments and the positions they occupy on an *L*-moment ratio diagram. Cong et al. (1993) constructed statistics based on the scatter of the (t_3, t_4) points for different sites on a *L*-moment ratio diagram about the τ_3–τ_4 relations of different three-parameter distributions. Their aim was to choose the distribution that gives the best fit to the data in this sense. As noted above, we believe that this should not be the sole aim of the analysis.

We prefer an alternative approach that works directly with the regional average *L*-moment statistics. It is described in the next section.

5.2 A goodness-of-fit measure

5.2.1 Aim

Given a set of sites that constitute a homogeneous region, the aim is to test whether a given distribution fits the data acceptably closely. A related aim is to choose, from a number of candidate distributions, the one that gives the best fit to the data.

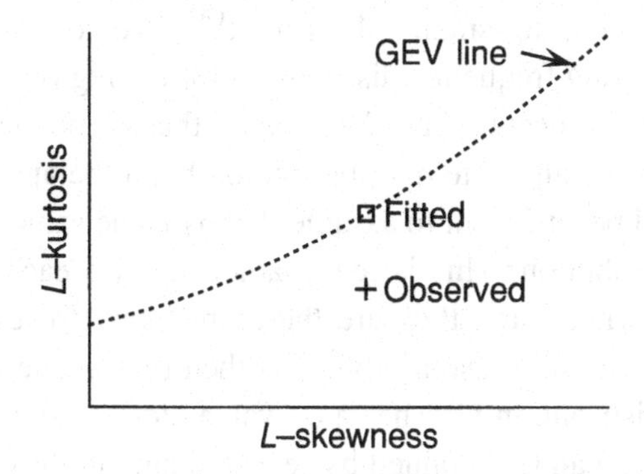

Fig. 5.1. Definition sketch for goodness of fit.

5.2.2 Heuristic description

Assume that the region is acceptably close to homogeneous. Choice of a distribution for regions that are not homogeneous is discussed in Section 5.3. The L-moment ratios of the sites in a homogeneous region are well summarized by the regional average; the scatter of the individual sites' L-moment ratios about the regional average represents no more than sampling variability. In most cases, the distribution being tested will have location and scale parameters that can be chosen to match the regional average mean and L-CV. The goodness of fit will therefore be judged by how well the L-skewness and L-kurtosis of the fitted distribution match the regional average L-skewness and L-kurtosis of the observed data. Fifth- or higher-order L-moments could in principle be used too, but we have not found it necessary to do so.

To obtain a goodness-of-fit measure, we argue as follows. Assume for convenience that the candidate distribution is generalized extreme-value (GEV), which has three parameters, and that sample L-skewness and L-kurtosis are exactly unbiased. The GEV distribution fitted by the method of L-moments has L-skewness equal to the regional average L-skewness. We therefore judge the quality of fit by the difference between the L-kurtosis τ_4^{GEV} of the fitted GEV distribution and the regional average L-kurtosis t_4^{R}. See Figure 5.1. To assess the significance of this difference, we compare it with the sampling variability of t_4^{R}. Let σ_4 denote the standard deviation of t_4^{R}, which can be obtained by repeated simulation of a homogeneous region whose sites have a GEV frequency distribution and record lengths the same as those of the observed data. Then

$$Z^{\mathrm{GEV}} = (t_4^{\mathrm{R}} - \tau_4^{\mathrm{GEV}})/\sigma_4 \tag{5.3}$$

is a goodness-of-fit measure; small values of Z^{GEV} are consistent with the GEV being the true underlying frequency distribution for the region.

A difficulty with the procedure just described is that a separate set of simulations must be made for each candidate distribution to obtain the appropriate σ_4 values. In practice, it should be sufficient to assume that σ_4 is the same for each candidate three-parameter distribution. This is reasonable, because each fitted distribution has the same *L*-skewness, and they are therefore likely to resemble each other to a large extent. Given this assumption, it is then reasonable to assume that the best-fitting kappa distribution also has a σ_4 value close to those of the candidate distributions. Thus σ_4 can be obtained by repeated simulations of a kappa region. These simulations can be the ones used in the calculation of the heterogeneity measure described in Section 4.3.

We have so far taken the sample *L*-moments t_3 and t_4 to be exactly unbiased. This is a very good approximation for t_3 but is not so good for t_4 when record lengths are short ($n_i \le 20$) or the population *L*-skewness is large ($\tau_3 \ge 0.4$). To overcome this problem, a bias correction for t_4 is used. Compare the fitted *L*-kurtosis τ_4^{GEV} not with the regional average t_4^{R} itself but with the bias-corrected version $t_4^{\mathrm{R}} - B_4$ where B_4 is the bias in the regional average *L*-kurtosis for regions with the same number of sites and the same record lengths as the observed data. This bias can be obtained from the same simulations as those used to obtain σ_4.

The foregoing description has dealt exclusively with three-parameter distributions. Two-parameter distributions can be treated similarly, but problems arise at the stage corresponding to the estimation of σ_4. The matter is discussed further in Subsection 5.2.6.

5.2.3 Formal definition

Suppose that the region has N sites, with site i having record length n_i and sample *L*-moment ratios $t^{(i)}$, $t_3^{(i)}$, and $t_4^{(i)}$. Denote by t^{R}, t_3^{R}, and t_4^{R} the regional average *L*-CV, *L*-skewness, and *L*-kurtosis, weighted proportionally to the sites' record length, as in Eq. (4.3).

Assemble a set of candidate three-parameter distributions. Reasonable possibilities include the generalized logistic (GLO), generalized extreme-value (GEV), generalized Pareto, lognormal, and Pearson type III.

Fit each distribution to the regional average *L*-moments 1, t^{R}, and t_3^{R}. Denote by τ_4^{DIST} the *L*-kurtosis of the fitted distribution, where DIST can be any of GLO, GEV, etc.

Fit a kappa distribution to the regional average *L*-moment ratios 1, t^{R}, t_3^{R}, and t_4^{R}.

Simulate a large number, N_{sim}, of realizations of a region with N sites, each having this kappa distribution as its frequency distribution. The simulated regions are homogeneous and have no cross-correlation or serial correlation; sites have the same record lengths as their real-world counterparts. The fitting of a kappa distribution and simulation of kappa regions can use the same computations as for the heterogeneity measure described in Section 4.3. For the mth simulated region, calculate the regional average L-skewness $t_3^{[m]}$ and L-kurtosis $t_4^{[m]}$.

Calculate the bias of t_4^{R},

$$B_4 = N_{\mathrm{sim}}^{-1} \sum_{m=1}^{N_{\mathrm{sim}}} (t_4^{[m]} - t_4^{\mathrm{R}}), \tag{5.4}$$

the standard deviation of t_4^{R},

$$\sigma_4 = \left[(N_{\mathrm{sim}} - 1)^{-1} \left\{ \sum_{m=1}^{N_{\mathrm{sim}}} (t_4^{[m]} - t_4^{\mathrm{R}})^2 - N_{\mathrm{sim}} B_4^2 \right\} \right]^{1/2}, \tag{5.5}$$

and, for each distribution, the goodness-of-fit measure

$$Z^{\mathrm{DIST}} = (\tau_4^{\mathrm{DIST}} - t_4^{\mathrm{R}} + B_4)/\sigma_4 . \tag{5.6}$$

Declare the fit to be adequate if Z^{DIST} is sufficiently close to zero, a reasonable criterion being $|Z^{\mathrm{DIST}}| \le 1.64$.

5.2.4 Performance

The performance of Z as a goodness-of-fit measure was assessed by means of Monte Carlo simulation experiments. Data were simulated from homogeneous regions with one of four three-parameter frequency distributions: generalized logistic (GLO), generalized extreme-value (GEV), lognormal (LN3), or Pearson type III (PE3). Simulations were made of data from each region; 1000 replications of each region were simulated. Each of these four distributions was also fitted to each region's data, and counts were kept of the number of times that each distribution was accepted as giving an adequate fit to the data, that is, $|Z| \le 1.64$, and of the number of times that each distribution was chosen as giving the best fit among the four fitted distributions in the sense of giving the smallest value of $|Z|$. Z statistics were calculated with $N_{\mathrm{sim}} = 500$.

Simulation results are given in Table 5.2. From the construction of Z, the true distribution of the region should be accepted about 90% of the time. This

Table 5.2. *Simulation results for the goodness of fit measure Z.*

			% Accepted				% Chosen			
τ	τ_3	Region	GLO	GEV	LN3	PE3	GLO	GEV	LN3	PE3
0.10	0.05	GLO	74	3	7	6	91	0	9	0
		GEV	2	90	82	85	1	61	23	15
		LN3	7	82	89	89	5	36	43	15
		PE3	6	85	89	90	4	39	41	15
0.20	0.20	GLO	80	25	16	2	89	10	1	0
		GEV	34	95	89	53	15	53	20	11
		LN3	15	90	93	71	4	39	31	26
		PE3	1	52	71	90	0	6	23	70
0.30	0.30	GLO	86	54	18	0	85	14	1	0
		GEV	74	95	69	8	36	47	15	0
		LN3	13	85	95	36	4	35	50	11
		PE3	0	5	41	93	0	0	14	85

Note: Simulations are of homogeneous regions with specified values of *L*-CV, τ, and *L*-skewness, τ_3. Each region has 21 sites and record length 30 at each site. "Region" is the true distribution used in the simulations. Here, "% Accepted" is the percentage of the simulations in which a candidate distribution gave an acceptable fit ($|Z| \leq 1.64$); "% Chosen" is the percentage of the simulations in which a distribution was chosen as the best of the four candidates in the sense of giving the smallest value of $|Z|$.

is approximately true for all parent distributions except the generalized logistic, which is accepted less often. It is not clear why this should be so; it may reflect a tendency for σ_4 to underestimate the true variance of the regional average *L*-kurtosis for generalized logistic regions. The amount by which these numbers exceed the other entries in the "% Accepted" columns of Table 5.2 measures the ability of Z to distinguish between different distributions. This is achieved fairly well for the generalized logistic distribution and, when τ_3 is relatively high, for the Pearson type III distribution, but in other cases the distributions are hard to distinguish. This reflects the similarity of the quantiles of the generalized extreme-value, lognormal, and Pearson type III distributions when τ_3 is small. In particular, the lognormal and Pearson type III distributions both tend to the Normal distribution when τ_3 tends to zero and are very similar when $\tau_3 = 0.05$. This explains the similarity of the corresponding rows of Table 5.2. The entries in the "% Chosen" columns show how well the Z statistic can be used to identify the correct distribution from among the four candidates. Again, this can be achieved particularly well for the generalized logistic distribution and for the Pearson type III distribution with high τ_3.

5.2.5 Notes

The criterion $|Z| \leq 1.64$ is somewhat arbitrary. The Z statistic has the form of a significance test of goodness of fit and has approximately a standard Normal distribution under suitable assumptions. The criterion $|Z| \leq 1.64$ then corresponds to acceptance of the hypothesized distribution at a confidence level of 90%. However, the assumptions necessary for Z to be standard Normal include two that are unlikely to be exactly satisfied in practice: that the region be exactly homogeneous and that it have no intersite dependence. Thus the criterion is a rough indicator of goodness of fit and is not recommended as a formal test.

The criterion $|Z| \leq 1.64$ is particularly unreliable if serial correlation or cross-correlation is present in the data. Correlation tends to increase the variability of t_4^R, and because there is no correlation in the simulated kappa region, the resulting estimate of σ_4 is too small and the Z values are too large. Thus a false indication of poor fit may be given. To overcome this problem, it is possible to generate simulated data that are correlated, though this would require much more computing time.

5.2.6 Goodness of fit for two-parameter distributions

A two-parameter distribution with location and scale parameters has fixed τ_3 and τ_4. Testing goodness of fit can be based on comparison of these τ_3 and τ_4 values with the regional average t_3^R and t_4^R. To construct a statistic analogous to Z, we need both the individual and joint sampling variability of t_3^R and t_4^R, in the form of the covariance matrix of t_3^R and t_4^R. Three plausible methods of obtaining this covariance matrix all have disadvantages.

(i) Estimate the covariance matrix by simulation of a kappa region, exactly as described above for three-parameter distributions. The assumption that σ_4 and the other terms in the covariance matrix are approximately the same for this kappa distribution as for the candidate distribution is now suspect, because the regional averages t_3^R and t_4^R may both be quite different from those of the candidate distribution. Thus the estimated covariance matrix may be unreliable.

(ii) Estimate the covariance matrix by simulation of a region whose frequency distribution is the candidate distribution. This requires more computing time, because the simulation procedure must be repeated for each candidate two-parameter distribution.

(iii) Calculate the covariance matrix from asymptotic theory for the distribution of t_3 and t_4 from the candidate distribution. The use of asymptotics may be a problem, because asymptotic approximations for distributions of L-moments are not always accurate in small samples ($n < 50$) – see, for example, Chowdhury et al. (1991, Table 1). Furthermore, it is not possible to obtain the asymptotic distributions for samples from serially correlated or cross-correlated data. Thus this approach is limited to situations in which it is adequate to obtain the covariance matrix of t_3^R and t_4^R for a region that has no serial or intersite dependence.

None of the foregoing methods is completely satisfactory. Method (ii) is a reasonable choice if sufficient computing time is available. Otherwise, we do not recommend the use of a goodness-of-fit measure for two-parameter distributions in regional frequency analysis.

5.3 Use of the goodness-of-fit measure

The procedure for a region that is acceptably homogeneous is as follows. Calculate Z for all candidate distributions. Flag as "acceptable" all distributions for which $|Z| \leq 1.64$. Calculate growth curves for the acceptable distributions. If these growth curves are approximately equal, for the scientific purposes of the application under consideration, then any of the acceptable distributions is adequate. To guard against the possibility that the region was misspecified, it is safest to choose from among the acceptable distributions the one that is most robust to such misspecification. If the growth curves are not approximately equal, there is a problem of scarcity of data; two models display differences that are statistically insignificant but operationally important. In this case, in which it has not been possible to confidently identify the best model, robustness becomes particularly important. Rather than choose a three-parameter distribution, it may be better to use the four-parameter kappa or five-parameter Wakeby distributions, which are more robust to misspecification of the frequency distribution of a homogeneous region – see Section 7.5.8.

It may happen that none of the candidate distributions is accepted by the Z criterion. This sometimes occurs when the number of sites in the region or the at-site record lengths are large. In these circumstances, σ_4 is small and Z can be large even if the regional average L-skewness and L-kurtosis are fairly close to those of one of the candidate distributions. If the regional average (t_3^R, t_4^R) point falls between two distributions (or among three or more distributions) whose growth curves are approximately equal, for the scientific purposes of the application under consideration, then there is a problem of superabundance of data; two models display differences that are statistically significant but operationally unimportant. In this case, it is reasonable to reclassify any of the operationally equivalent distributions as giving an acceptable fit to the data. Sometimes the regional average point does

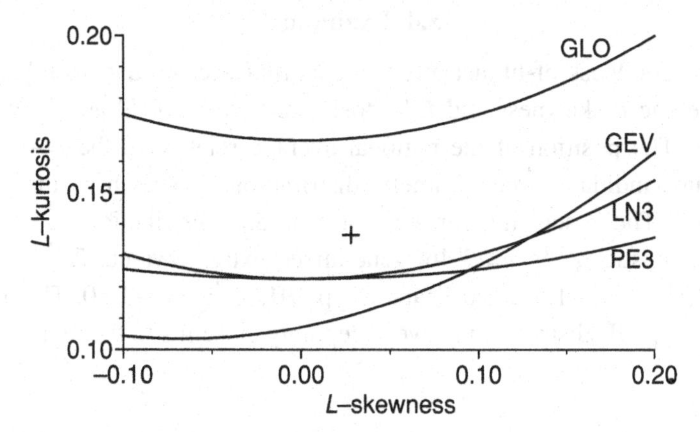

Fig. 5.2. Average *L*-moments of the North Cascades data (+), with *L*-skewness–*L*-kurtosis relationships for four three-parameter distributions.

not lie between two operationally equivalent distributions. For example, it may lie above the generalized-logistic line. In these cases, no three-parameter distribution is acceptable, and a more general distribution such as the kappa or Wakeby should be used.

If the region is not acceptably homogeneous, there is no reason to suppose that a single distribution will give a good fit to every site's data. Nonetheless, fitting a single distribution can still yield much more accurate quantile estimates, considered on the average over all sites, than fitting separate distributions at each site – see Chapter 7, and in particular Sections 7.5.3 and 7.5.7 – 7.5.10. The choice of distribution should be influenced by considerations of robustness. It is particularly important to use a distribution that is robust to moderate heterogeneity in the at-site frequency distributions. The kappa and Wakeby distributions are widely recommendable choices – see Sections 7.5.8–7.5.9.

When the region is heterogeneous, it is possible that a test that makes use of the at-site *L*-moments might enable better discrimination between distributions. The regional average gives a sufficient summary of the data when the region is homogeneous, but this is no longer the case for a heterogeneous region. However, for heterogeneous regions we consider it more important that the chosen distribution be robust to heterogeneity than that it achieve the ultimate quality of fit. We therefore tend to prefer the Wakeby distribution for heterogeneous regions.

In a large investigation there may be many regions, and the choice of frequency distribution for one region may affect the others. If one distribution gives an acceptable fit for all or most of the regions, then it is reasonable to use this distribution for all regions even though it may not be the best for each region individually.

5.4 Example

We apply the goodness-of-fit measure to the North Cascades data in Table 3.4. The regional average L-skewness and L-kurtosis are $t_3^R = 0.0279$ and $t_4^R = 0.1366$, respectively. The position of the regional average relative to the τ_3–τ_4 relationships of four candidate three-parameter distributions is shown by the + symbol on Figure 5.2. The Z statistics for the four candidate distributions are as follows: generalized logistic, $Z^{\mathrm{GLO}} = 3.46$; generalized extreme-value, $Z^{\mathrm{GEV}} = -2.94$; lognormal, $Z^{\mathrm{LN3}} = -1.51$; and Pearson type III, $Z^{\mathrm{PE3}} = -1.60$. The lognormal and Pearson type III distributions give acceptably close fits to the regional average L-moments.

6
Estimation of the frequency distribution

6.1 Estimation for a homogeneous region

After successful use of the methods described in Chapters 3–5, the sites at which data are available for regional frequency analysis will have been assigned to regions that are nearly homogeneous, that is, the frequency distributions at the sites in a region are approximately identical apart from a scale factor, and a probability distribution will have been chosen for fitting to each region's data. The relationship between the frequency distributions at different sites is the justification for regional frequency analysis. It is this relationship that enables more accurate estimates of the distribution's parameters and quantiles to be obtained by combining the data from different sites than could be achieved by fitting distributions to each site's data separately.

Several methods of fitting a distribution to data from a homogeneous region are possible. To describe them, we recapitulate the notation of Section 1.3. Suppose that the region has N sites, with site i having sample size n_i and observed data Q_{ij}, $j = 1, \ldots, n_i$. Let $Q_i(F)$, $0 < F < 1$, be the quantile function of the frequency distribution at site i. For a homogeneous region we have

$$Q_i(F) = \mu_i q(F), \qquad i = 1, \ldots, N, \tag{6.1}$$

where μ_i is the site-dependent scale factor, the index flood. Let $\hat{\mu}_i$ be the estimate of the scale factor at site i. The dimensionless rescaled data are $q_{ij} = Q_{ij}/\hat{\mu}_i$, $j = 1, \ldots, n_i, i = 1, \ldots, N$.

The station-year method combines the rescaled data from all sites into a single sample and fits a distribution by treating the combined sample as a single random sample. The method is now rarely used, because in many cases it is not appropriate to treat the rescaled data as a single random sample. When the estimates $\hat{\mu}_i$ have different accuracies, as is the case when they are calculated from the at-site data and the sites have different record lengths, the rescaled data from different sites will not be identically distributed.

An approach based on maximum-likelihood estimation treats Eq. (6.1) as a statistical model that is completely specified by the N scale factors μ_i, $i = 1, \ldots, N$, and the p unknown parameters of the regional growth curve $q(F)$. These $N + p$ parameters can be estimated by the method of maximum likelihood, iterative methods usually being necessary to find the maximum of the likelihood function. This method has been used by Boes, Heo, and Salas (1989) and Buishand (1989). It can also be used when the scale factors are not independent parameters but are themselves modeled as functions of exogenous variables; that is, $\mu_i = h(\mathbf{z}_i; \omega)$, where $\mathbf{z}_i$ is a vector of site characteristics for site i and ω is a vector of unknown parameters. Examples of the use of this approach include Moore (1987) and Smith (1989).

The index-flood procedure described in Section 1.3 uses summary statistics of the data at each site and combines them by averaging to form the regional estimates defined in Eq. (1.5). When the summary statistics are the L-moment ratios of the at-site data, we call the resulting procedure the "regional L-moment algorithm." It is fully described in Section 6.2. The index-flood procedure has no theoretical superiority to the maximum-likelihood approach but is an intuitively reasonable way of combining the information from different sites. The calculations that it requires tend to be simpler than those of maximum-likelihood estimation, because they do not involve the entire set of regional data simultaneously. The regional average L-moment ratios calculated in the regional L-moment algorithm are themselves useful as a summary of the salient features of the regional data set.

We regard the index-flood procedure implemented in the regional L-moment algorithm as a convenient and efficient method of estimating a regional frequency distribution, and we concentrate on it in the next two chapters. This chapter describes how the regional L-moment algorithm is used to estimate quantiles of the regional frequency distribution and how the accuracy of the estimates for a particular data set can be assessed. Chapter 7 is a more general survey of the accuracy of quantile estimates obtained by applying the regional L-moment algorithm to data from a wide range of homogeneous and heterogeneous regions, and includes comparisons of the accuracy of regional and at-site estimation.

The methods described above assume that there is no dependence between observations at different sites and no serial dependence between observations at the same site. Dependence could in principle be built into the statistical model used in the maximum-likelihood approach, but in practice a suitable dependence structure would be very complicated both to specify and to estimate. Provided that dependence is not too strong, as is the case in many realistic situations, the accuracy of the quantile estimates obtained by each of the methods should not be greatly affected. The robustness of the regional L-moment algorithm to dependence between observations and to the breaking of the other assumptions of the index-flood procedure, listed in Section 1.3, is among the subjects explored in Chapter 7.

6.2 The regional L-moment algorithm

6.2.1 *Aim*

The aim is to fit, to the data from the sites in a homogeneous region, a single frequency distribution (the regional frequency distribution) that describes the distribution of the observations at each site after scaling by the at-site scaling factor (index flood). This distribution is then scaled appropriately at each site in order to estimate quantiles of the at-site frequency distributions.

6.2.2 *Heuristic description*

The distribution is fitted by the method of L-moments; its parameters are estimated by equating the population L-moments of the distribution to the sample L-moments calculated from the data. Assuming the region to be homogeneous, sample L-moment ratios calculated from the rescaled data for different sites can be combined to give regional average L-moment ratios. To allow for the greater variability of L-moment ratios in small samples, averages are weighted proportionally to the sites' record lengths.

For simplicity we assume that the index flood is the mean of the frequency distribution at each site and that it is estimated by the sample mean of the at-site data. Then the mean of the rescaled data is 1 for each site, and so the regional average of these means is 1. Furthermore, for each site the sample L-moment ratios t and t_r, $r \geq 3$, are the same whether calculated from the rescaled data $\{q_{ij}\}$ or the original data $\{Q_{ij}\}$. The explicit computation of the rescaled data is therefore unnecessary.

6.2.3 *Formal definition*

As above, we assume that the index flood is the mean of the frequency distribution at each site and that it is estimated at site i by the sample mean of the at-site data.

Suppose that the region has N sites, with site i having record length n_i, sample mean $\ell_1^{(i)}$, and sample L-moment ratios $t^{(i)}, t_3^{(i)}, t_4^{(i)}, \ldots$. Denote by $t^{\mathrm{R}}, t_3^{\mathrm{R}}, t_4^{\mathrm{R}}, \ldots,$ the regional average L-moment ratios, weighted proportionally to the sites' record length:

$$t^{\mathrm{R}} = \sum_{i=1}^{N} n_i t^{(i)} \Big/ \sum_{i=1}^{N} n_i \,, \tag{6.2}$$

$$t_r^{\mathrm{R}} = \sum_{i=1}^{N} n_i t_r^{(i)} \Big/ \sum_{i=1}^{N} n_i \,, \qquad r = 3, 4, \ldots. \tag{6.3}$$

Set the regional average mean to 1, that is, $\ell_1^{\mathrm{R}} = 1$.

Fit the distribution by equating its L-moment ratios λ_1, τ, τ_3, τ_4, ... , to the regional average L-moment ratios ℓ_1^{R}, t^{R}, t_3^{R}, t_4^{R}, ..., calculated above. Denote by $\hat{q}(.)$ the quantile function of the fitted regional frequency distribution.

The quantile estimates at site i are obtained by combining the estimates of μ_i and $q(.)$. The estimate of the quantile with nonexceedance probability F is

$$\hat{Q}_i(F) = \ell_1^{(i)} \hat{q}(F). \tag{6.4}$$

6.2.4 *Variants of the regional L-moment algorithm*

Several variants of the basic regional L-moment algorithm may be entertained.

The procedure described in the previous subsection supposes that the index flood is the mean of the at-site frequency distribution. If instead it is some other quantity, such as the median or another quantile of the distribution, the fitted regional frequency distribution should be rescaled so that the index flood of this distribution is 1. Let $\hat{q}(.)$ now denote the quantile function of this rescaled distribution, and let $\hat{\mu}_i$ be the estimate of the index flood for site i. Quantile estimates are then given by Eq. (6.4) but with $\hat{\mu}_i$ replacing $\ell_1^{(i)}$, that is,

$$\hat{Q}_i(F) = \hat{\mu}_i \hat{q}(F). \tag{6.5}$$

The calculation of regional averages by weighting the sites proportionally to their record lengths is not essential. If the region is exactly homogeneous, then to a good approximation the variance of $t_r^{(i)}$ is proportional to n_i^{-1}, and in this case weighting the sites proportionally to their record lengths minimizes the variance of the regional average t_r^{R}. If the region is heterogeneous, it is possible that weighting proportionally to record length may give undue influence to sites that have frequency distributions markedly different from the region as a whole and that also have long records. For this reason, Jin and Stedinger (1989) preferred to use an unweighted average. An alternative weighting, giving less weight to the sites with the longest records, has been suggested by Stedinger, Vogel, and Foufoula-Georgiou (1992).

The L-moment ratios used in the algorithm can be computed using the plotting-position estimators described in Section 2.8 rather than the "unbiased" estimators. Use of plotting-position estimators sometimes gives an improvement in the accuracy of quantile estimates in the extreme upper tail of frequency distributions with high skewness. An example is given in Section 7.5.2 (see also Hosking and Wallis, 1995). However, as noted in Section 2.8, "unbiased" estimators are superior for most purposes and we prefer to work with them throughout.

Regional averages can be computed for the L-moments rather than the L-moment ratios of the rescaled data. In terms of the original data this means that the regional average L-moments would be defined by

$$(\ell_r^{\mathrm{R}})^* = \sum_{i=1}^{N} n_i \frac{\ell_r^{(i)}}{\ell_1^{(i)}} \bigg/ \sum_{i=1}^{N} n_i \,. \tag{6.6}$$

This is the procedure described by Stedinger et al. (1992, Section 18.5.1). This yields $(\ell_1^{\mathrm{R}})^* = 1$, and the same value as in Eq. (6.2) for the regional average L-CV, but is equivalent to calculating the regional average third- and higher-order L-moment ratios by

$$(t_r^{\mathrm{R}})^* = \sum_{i=1}^{N} n_i \, t^{(i)} \, t_r^{(i)} \bigg/ \sum_{i=1}^{N} n_i \, t^{(i)} \,. \tag{6.7}$$

Averaging L-moment ratios rather than L-moments is preferable, because it yields slightly more accurate quantile estimates in almost all of the cases that we have investigated. An example is given in Section 7.5.1.

As a historical note, some of the variants mentioned above were used in the earliest applications of probability weighted moments to regional frequency analysis. Wallis (1981) obtained "regionally estimated probability weighted moments" by using an unweighted average of plotting-position estimators of probability weighted moments – equivalent to L-moments, not L-moment ratios – calculated from the rescaled data. Wallis (1982) and Hosking et al. (1985a), using the "regional GEV/PWM" and "regional WAK/PWM" algorithms, incorporated weighting proportional to record length but still used plotting-position estimators, rather than "unbiased" estimators, and probability weighted moments, rather than L-moment ratios, of the rescaled data.

6.3 Example: Estimation of the regional frequency distribution

For the North Cascades data of Table 3.4, the goodness-of-fit analysis in Section 5.4 found that the lognormal and Pearson type III distributions give acceptably close

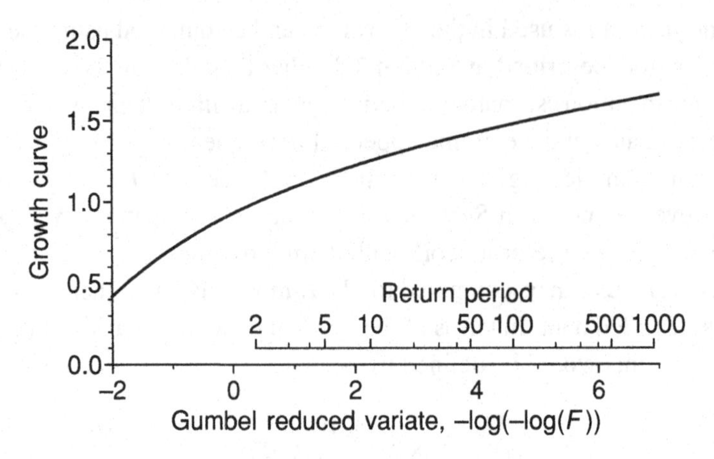

Fig. 6.1. Regional quantile function fitted to the North Cascades data.

fits to the regional average *L*-moments. The regional average mean, *L*-CV, and *L*-skewness are

$$\ell_1^R = 1, \qquad t^R = 0.1103, \qquad t_3^R = 0.0279. \tag{6.8}$$

The parameters of the fitted distributions, obtained from Eqs. (A.74) and (A.75) and Eqs. (A.90) and (A.92), are as follows:

$$\text{lognormal:} \quad \xi = 0.9944, \quad \alpha = 0.1952, \quad k = -0.0571; \tag{6.9}$$

$$\text{Pearson type III:} \quad \mu = 1.0000, \quad \sigma = 0.1957, \quad \gamma = 0.1626. \tag{6.10}$$

The growth curves for these two distributions are almost identical throughout the range of quantiles from 0.01 to 0.999, so either distribution would be an appropriate choice for this region. The growth curves are plotted, as though on extreme-value probability paper, in Figure 6.1; the two curves are indistinguishable.

6.4 Assessment of the accuracy of estimated quantiles

Results obtained by statistical analysis are inherently uncertain, and for the results to be maximally useful some assessment of the magnitude of uncertainty should be made. In traditional statistics this is achieved by the construction of confidence intervals for estimated parameters and quantiles, usually assuming that all of the statistical model's assumptions are satisfied. In regional frequency analysis using the regional *L*-moment algorithm it is similarly possible to construct confidence intervals for estimation in homogeneous regions, at least as a large-sample approximation when sample L-moments may be taken to be Normally distributed. Analogous results can be obtained for other methods of regional frequency analysis, but the calculated accuracies often merely reflect the stringency of the assumptions made by different methods, as Rosbjerg and Madsen (1995) found; when each method's accuracy is calculated on the basis that the method's assumptions are satisfied, then methods that make strictest assumptions also claim to give the most accurate results.

Such confidence intervals are of limited utility in practice, because we can rarely be sure that the "correct" model was used, that is, in the case of the regional L-moment algorithm, that the data satisfy all of the assumptions, listed in Section 1.3, that underlie the index-flood procedure. Indeed, one of the strengths of regional frequency analysis using the regional L-moment algorithm is that it is useful even when not all of its assumptions are satisfied. A realistic assessment of the accuracy of estimates should therefore take into account the possibility of heterogeneity in the region, misspecification of the frequency distribution, and statistical dependence between observations at different sites, to an extent that is consistent with the data.

A reasonable approach is to estimate the accuracy of estimated quantiles by Monte Carlo simulation. The simulations should be matched to the particular characteristics of the data from which the estimates are calculated. The region used as the basis for simulation should be chosen to have the same number of sites, record length at each site, and regional average L-moment ratios as the actual data. As noted above, it will often be appropriate for the simulated region to include heterogeneity, misspecification, and intersite dependence or some combination thereof.

The L-moment ratios at the individual sites should be chosen to yield a region whose heterogeneity is consistent with the heterogeneity measures calculated from the data. Some preliminary simulations may be needed to establish how much variation among the at-site L-moment ratios is needed to yield the observed values of the heterogeneity measures. Some arbitrariness is inevitable here, because many different patterns of variation in the at-site L-moment ratios may be consistent with the observed values of the heterogeneity measures.

One important point is that the between-site variation in population L-moment ratios for the simulated region should always be less than that of the sample L-moment ratios of the actual data, because sampling variability causes sample L-moment ratios to be much more widely scattered than the corresponding population L-moment ratios. Illustrations are given in the example in Section 6.5 and in Figure 7.2. In particular, it is incorrect to use the observed sample L-moment ratios as the population L-moment ratios of the simulated region, because this would yield a simulated region that has much more heterogeneity than the actual data.

The frequency distributions at individual sites should be chosen to be consistent with the goodness-of-fit measures obtained for the data. If several distributions appear to fit well, then any of them is a plausible candidate for use in the simulations; if no distribution fits well, then some flexible distribution such as the Wakeby or kappa can be used.

If intersite dependence is thought to be a problem, then it can be included in the simulations by modifying the simulation algorithm. A particularly convenient form of dependence structure arises from assuming that if each site's frequency distribution were transformed to the Normal distribution – call this transformation $\mathcal{T}$ – then the joint distribution for all N sites would be multivariate Normal. This is a realistic dependence structure for at least some kinds of environmental data. For example, Hosking and Wallis (1987b) found this assumption to be fairly well supported for British annual maximum streamflow series. Data generation then involves the following steps: generate the matrix $\mathbf{R}$ of intersite correlations; generate a random vector $\mathbf{y}$ having a multivariate Normal distribution with covariance matrix $\mathbf{R}$; and apply the inverse of transformation $\mathcal{T}$ to obtain data with the required marginal distribution.

A simulation algorithm that implements this procedure is given in Table 6.1. Step 3.1 of the algorithm shows the modifications required when intersite dependence is incorporated into the simulations.

The correlation matrix $\mathbf{R}$ used in simulation of correlated data should be chosen to be consistent with the correlation patterns in the data. When there is no particular pattern of correlation among the sites in the region, it is reasonable to take the sites to be equicorrelated, that is, the correlation between sites i and j is $\rho_{ij} = \rho$ for $i \neq j$ and the matrix $\mathbf{R}$ has the form

$$\mathbf{R} = \begin{bmatrix} 1 & \rho & \rho & \ldots & \rho \\ \rho & 1 & \rho & \ldots & \rho \\ \rho & \rho & 1 & \ldots & \rho \\ \vdots & \vdots & & \ddots & \\ \rho & \rho & & & 1 \end{bmatrix} . \tag{6.11}$$

The quantity ρ may be estimated by the average cross-correlation of the data at all pairs of sites. Let Q_{ik} be the data value for site i at time point k. The sample correlation between sites i and j is given by

$$r_{ij} = \frac{\sum_k (Q_{ik} - \bar{Q}_i)(Q_{jk} - \bar{Q}_j)}{\left\{\sum_k (Q_{ik} - \bar{Q}_i)^2 \sum_k (Q_{jk} - \bar{Q}_j)^2\right\}^{1/2}} , \tag{6.12}$$

Table 6.1. *Algorithm for simulation of the regional L-moment algorithm.*

1. Specify N and for each of the N sites its record length n_i and the L-moments of its frequency distribution.
2. Calculate the parameters of the at-site frequency distributions given their L-moment ratios.
3. For each of M repetitions of the simulation procedure, carry out the following steps.
   - 3.1. Generate sample data for each site. If there is no intersite dependence, this simply requires the generation of a random sample of size n_i from the frequency distribution for site i, $i = 1, \ldots, n$. If intersite dependence is included in the simulations, the following procedure can be used.
     - 3.1.1. Let $n_0 = \max n_i$, the largest of the at-site record lengths. For each time point $k = 1, \ldots, n_0$, generate a realization of a random vector $\mathbf{y}_k$ with elements y_{ik}, $i = 1, \ldots, N$, that has a multivariate Normal distribution with mean vector zero and covariance matrix $\mathbf{R}$.
     - 3.1.2. Transform each y_{ik}, $k = 1, \ldots, n_i$, $i = 1, \ldots, N$, to the required marginal distribution, that is, calculate the data values $Q_{ik} = Q_i\big(\Phi(y_{ik})\big)$, where Q_i is the quantile function for site i and Φ is the cumulative distribution function of the standard Normal distribution.
   - 3.2. Apply the regional L-moment algorithm to the sample of regional data. This involves the following steps:
     - 3.2.1. Calculate at-site L-moment ratios and regional average L-moment ratios;
     - 3.2.2. Fit the chosen distribution;
     - 3.2.3. Calculate estimates of the regional growth curve and at-site quantiles.
   - 3.3. Calculate the relative error of the estimated regional growth curve and at-site quantiles, and accumulate the sums needed to calculate overall accuracy measures.
4. Calculate overall measures of the accuracy of the estimated quantiles and regional growth curve.

where

$$\bar{Q}_i = n_{ij}^{-1} \sum_k Q_{ik} \, ; \tag{6.13}$$

the sums over k extend over all time points for which sites i and j both have data, and n_{ij} is the number of such time points. The average intersite correlation is given by

$$\bar{r} = \{\tfrac{1}{2} N(N-1)\}^{-1} \sum_{1 \le i < j \le N} \sum r_{ij} \, . \tag{6.14}$$

More elaborate correlation structures may be used, if justified by physical reasoning about the similarity of different sites. One possibility, used in the simulations described in Section 7.5.6, is to permit ρ_{ij} to be a function of the distance between sites i and j. However, the principal effect of intersite dependence, a general increase in the RMSE of quantile and growth curve estimates, should usually be sufficiently well captured by the use of equicorrelated data, with correlation matrix of the form (6.11).

In the simulation procedure, quantile estimates are calculated for various nonexceedance probabilities. At the mth repetition, let the site-i quantile estimate for nonexceedance probability F be $\hat{Q}_i^{[m]}(F)$. The relative error of this estimate is $\{\hat{Q}_i^{[m]}(F) - Q_i(F)\}/Q_i(F)$. This quantity can be squared and averaged over all M repetitions to approximate the relative RMSE of the estimators. The relative RMSE is approximated, for large M, by

$$R_i(F) = \left[M^{-1} \sum_{m=1}^{M} \left\{ \frac{\hat{Q}_i^{[m]}(F) - Q_i(F)}{Q_i(F)} \right\}^2 \right]^{1/2} . \tag{6.15}$$

A summary of the accuracy of estimated quantiles over all of the sites in the region is given by the regional average relative RMSE of the estimated quantile,

$$R^{\mathrm{R}}(F) = N^{-1} \sum_{i=1}^{N} R_i(F) . \tag{6.16}$$

In addition to the overall accuracy measures for quantile estimates, Eqs. (6.15) and (6.16), analogous quantities can be calculated for the growth curve estimate. Let the growth curve for site i be $q_i(F)$, defined by

$$Q_i(F) = \mu_i q_i(F) . \tag{6.17}$$

The quantities $q_i(F)$ are needed for simulation of heterogeneous regions, whereas in a homogeneous region each $q_i(F)$ is equal to the regional growth curve $q(F)$. At the mth repetition, let the estimated regional growth curve be $\hat{q}^{[m]}(F)$. Accuracy measures for the estimated growth curve are defined by Eqs. (6.15) and (6.16) but with $Q_i(F)$ and $\hat{Q}_i^{[m]}(F)$ replaced by $q_i(F)$ and $\hat{q}^{[m]}(F)$, respectively. Accuracy measures for the growth curve are particularly relevant when only the growth curve estimate is of interest, as is the case when the index flood is estimated by methods not involving the at-site data or when quantiles are estimated for an ungaged site (Section 8.4).

Other useful quantities, particularly when the distribution of estimates is skew, are the empirical quantiles of the distribution of estimates. These can be obtained by calculating the ratio of estimated to true values, $\hat{Q}_i(F)/Q_i(F)$ for quantiles and $\hat{q}(F)/q_i(F)$ for the growth curve, averaging these values over the sites in the region, and accumulating over the different realizations a histogram of the values taken by the ratio. For example, for a particular nonexceedance probability F it may be found that 5% of the simulated values of $\hat{Q}(F)/Q(F)$ lie below some value $L_{.05}(F)$ whereas 5% lie above some value $U_{.05}(F)$. Then 90% of the distribution of $\hat{Q}(F)/Q(F)$ lies within the interval

$$L_{.05}(F) \leq \frac{\hat{Q}(F)}{Q(F)} \leq U_{.05}(F)\,, \tag{6.18}$$

and inverting this to express Q in terms of $\hat{Q}$ gives

$$\frac{\hat{Q}(F)}{U_{.05}(F)} \leq Q(F) \leq \frac{\hat{Q}(F)}{L_{.05}(F)}\,. \tag{6.19}$$

Expression (6.19) has the same form as a statistical confidence interval, but can be validly interpreted as one only if the distribution of $\hat{Q}(F)/Q(F)$ is independent of all of the parameters involved in the specification of the statistical model underlying the index-flood procedure; for the regional L-moment algorithm these parameters are the at-site means and the regional average L-moment ratios. In practice, independence does not hold and confidence statements are at best approximate. Nonetheless, the interval (6.19) should give a useful indication of the amount of variation between true and estimated quantities. We sometimes refer to the limits $\hat{Q}(F)/U_{.05}(F)$ and $\hat{Q}(F)/L_{.05}(F)$ in Eq. (6.19) as the "90% error bounds" for $\hat{Q}(F)$.

The error bounds given by Eq. (6.19) can be unhelpful in the lower tail of the distribution. If the fitted distribution can take negative values, it may happen that $L_{.05}(F)$ is very small, or even negative, leading to a very large or infinite upper bound in Eq. (6.19). In such cases, $R^{\mathrm{R}}(F)$, the regional average relative RMSE of the estimated quantiles, is a more informative measure of accuracy.

The simulation-based procedure leading to the bounds in Eq. (6.19) is less exact than the formal construction of confidence intervals but gives a reasonable estimate of the magnitude of the errors that can reasonably be expected to be present in the estimated quantiles and growth curve. The accuracy with which the error magnitude can be estimated depends on the number of repetitions, M, of the simulation procedure. Even $M = 100$ gives a useful indication of the magnitude of errors, but larger values of M will give more accurate error estimates. It is better to use $M = 1{,}000$ or even $M = 10{,}000$ when accurate judgements are required.

6.5 Example: Assessment of accuracy

In Section 6.3 we estimated the regional growth curve for the North Cascades annual precipitation totals summarized in Table 3.4. We now try to quantify the accuracy of the estimated regional growth curve. To decide on a suitable region for use in the simulation procedure, we assess its intersite dependence, at-site frequency distributions, and heterogeneity.

The region has a considerable amount of intersite dependence; correlations between sites are mostly between 0.2 and 0.9, with an average of 0.64. The simulation procedure therefore uses the algorithm of Table 6.1, incorporating intersite dependence. The correlation matrix used is Eq. (6.11), with $\rho = 0.64$.

The regional average L-moment ratios are 0.1103 (L-CV), 0.0279 (L-skewness), and 0.1366 (L-kurtosis). The goodness-of-fit measures calculated in Section 5.4 indicate that the lognormal and Pearson type III distributions are both consistent with the data. These distributions are very similar when their skewness is low, and it seems adequate to accept the lognormal as being the distribution for use in the simulations.

The heterogeneity measure Eq. (4.5) for the region is $H = 0.62$; corresponding measures based on V_2 and V_3 defined in Eqs. (4.6) and (4.7) are -1.49 and -2.37, respectively. These measures indicate no deviation from homogeneity, but it may be wise to be prepared for a little heterogeneity among the sites' L-CV values, consistent with the value of H. Simulations of correlated lognormal regions with record lengths the same as for the North Cascades data show that when at-site L-CVs vary over a range of 0.025, from 0.0978 to 0.1228, the average H value of simulated regions is 1.08. We use this range of L-CV in the simulation procedure, arbitrarily assigning the L-CV values in increasing order to the sites ordered as in Table 3.4. Note that the range of population L-CV values used in the simulations, 0.025, is considerably less than that of the sample L-CV values, given in Table 3.4, which have a range of 0.043.

The region used in the simulation procedure therefore contains 19 sites with record lengths as for the North Cascades data, the sites having lognormal frequency distributions with L-CV varying linearly from 0.0978 at site 1 to 0.1228 at site 19 and L-skewness 0.0279. 10,000 realizations of this region were made and the regional L-moment algorithm was used to fit a lognormal distribution to the data generated at each realization. This is the most computationally burdensome part of the analysis of the North Cascades data, requiring 240 seconds of CPU time on an IBM 390 (model 9021) mainframe. The regional average relative RMSE of the estimated growth curve was calculated from the simulations, and quantiles of the

Table 6.2. *Accuracy measures for estimated growth curve of the North Cascades data.*

F	$\hat{q}(F)$	RMSE	Error bounds	
0.01	0.569	0.073	0.523	0.625
0.1	0.753	0.028	0.731	0.776
0.5	0.994	0.005	0.985	1.004
0.9	1.254	0.017	1.230	1.278
0.99	1.480	0.033	1.418	1.540
0.999	1.654	0.047	1.547	1.755

Note: Region specification is described in text. Tabulated values are, for each nonexceedance probability F, the regional average relative RMSE of the estimated growth curve and the lower and upper 90% error bounds for the estimated growth curve, defined analogously to Eq. (6.19).

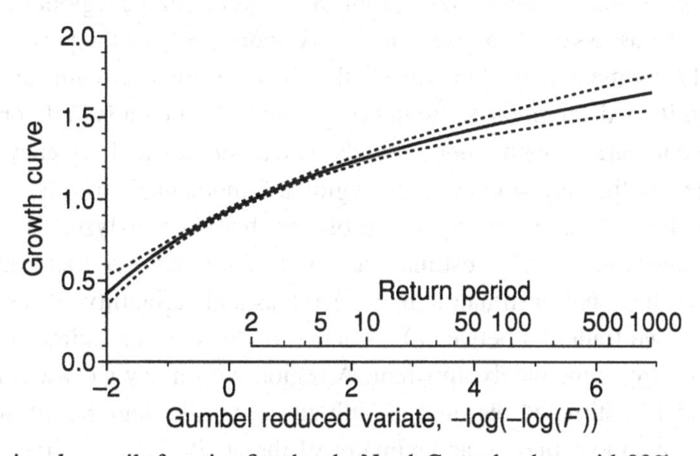

Fig. 6.2. Regional quantile function fitted to the North Cascades data, with 90% error bounds.

distribution of

$$N^{-1} \sum_{i=1}^{N} \hat{q}(F)/q_i(F), \tag{6.20}$$

the regional average of the ratio of the estimated to the true at-site growth curve, were computed from a histogram accumulated during the simulations. From these quantiles the 90% error bounds for the growth curve were computed analogously to Eq. (6.19). The results are given, for selected values of the nonexceedance probability F, in Table 6.2. Figure 6.2 shows the estimated growth curve, from Figure 6.1, together with its 90% error bounds.

7

Performance of the regional L-moment algorithm

7.1 Introduction

The methods described in Section 6.4 enable the accuracy of the regional L-moment algorithm to be assessed for a given region. A wider study of the performance of the regional L-moment algorithm is also valuable, to establish circumstances under which the utility of the regional L-moment algorithm is particularly high (or low) and to facilitate comparison with other methods such as single-site frequency analysis.

Assessment of the performance of the regional L-moment algorithm for different regions can also influence how a given set of data should be analyzed. The regional L-moment algorithm involves estimation of both the index flood and the growth curve, and each of these estimators may have bias and variability, as discussed in Section 2.2. Two tradeoffs between bias and variability of the estimated regional growth curve are immediately apparent. A region with many sites will have low variability of its estimated regional growth curve, but the regional growth curve is likely to have more bias as an estimator of the at-site growth curves, because exact homogeneity is less likely to hold for a region with many sites than for one with few. A fitted distribution with few parameters will yield less variability in the regional growth curve than a distribution with many parameters, but the distribution with few parameters is less likely to contain a good approximation to the true regional frequency distribution of the region, and its regional growth curve is therefore likely to be more biased. Some subjective judgement is required to achieve a suitable compromise between these conflicting criteria; the heterogeneity and goodness-of-fit measures described in Chapters 4 and 5 should be helpful.

Errors in quantile estimates obtained from the regional L-moment algorithm can be identified as arising from several different sources. A theoretical analysis is given in Section 7.2.

For more detailed results we rely on Monte Carlo simulation of some carefully chosen artificial regions. The simulation procedure is described in Section 7.3.

Results obtained by simulation are specific to the region used to generate the simulated data. It therefore requires many simulation experiments to determine the effects of the many factors that influence the accuracy of quantile estimates. Our approach has two main parts. In Section 7.4 we investigate the performance of the regional *L*-moment algorithm for a few regions whose specifications are representative of those that might be encountered in typical applications of regional frequency analysis. In Section 7.5 we explore in some detail the effect of different factors on the accuracy of the estimated growth curve and quantiles. Factors of interest include

- variants of the estimation procedure:
  regional averaging of τ and τ_3 versus λ_2/λ_1 and λ_3/λ_1;
  estimation based on "unbiased" versus plotting-position estimators; and
  regional versus at-site frequency analysis;
- aspects of the specification of the region:
  number of sites in the region; and
  record lengths at each site;
- violations of the assumptions of regional frequency analysis:
  intersite dependence;
  heterogeneity; and
  misspecification of the regional frequency distribution.

The other assumptions of regional frequency analysis listed in Section 1.3, that the data for any given site should be identically distributed and serially independent, are not explored as they affect both regional and at-site frequency analyses and do not affect our main concern, the utility of regional frequency analysis in comparison with at-site analysis.

Section 7.6 on page 141 summarizes the main conclusions of the simulation experiments and their consequences for data analysis. Readers not interested in the fine details of the results may prefer to turn directly to this section.

7.2 Theory

Several components that contribute to the error in quantile estimation with the regional *L*-moment algorithm can be identified from theoretical considerations. Consider the estimation procedure for a region of N sites that may be heterogeneous and may also be misspecified in that the "wrong" regional frequency distribution is fitted. The term "misspecified" is formally defined in this section.

Suppose that the frequency distribution at site i has quantile function $Q_i(F) = \mu_i q_i(F)$. Let $q^R(F)$ be the average growth curve, defined so as to give the best summary (in some sense) of the at-site growth curves $q_1(F), \ldots, q_N(F)$. We show below that a reasonable definition is $q^{\rm R}(F) = N / \sum\{1/q_i(F)\}$, the harmonic mean of the at-site growth curves. In a homogeneous region the at-site growth curves are identical: $q_i(F) = q^{\rm R}(F)$ for all i.

Suppose that the regional growth curve fitted to the data has the form $x(F;\boldsymbol{\theta})$, where $\boldsymbol{\theta} = [\theta_1 \ldots \theta_p]^{\rm T}$ is a vector of parameters. We say that the frequency distribution is *misspecified* if there is no value of $\boldsymbol{\theta}$ for which the functions $x(.;\boldsymbol{\theta})$ and $q^{\rm R}(.)$ are identical. Let $\boldsymbol{\theta}_0$ be the value of $\boldsymbol{\theta}$ that makes $x(.;\boldsymbol{\theta})$ as close as possible, in some sense, to $q^{\rm R}(.)$. For the regional L-moment algorithm, in which $\boldsymbol{\theta}$ is a vector of L-moment ratios, $\boldsymbol{\theta}_0$ would be the vector whose elements are the corresponding L-moment ratios of the distribution whose quantile function is $q^{\rm R}(F)$.

The index flood at site i is μ_i, estimated by $\hat{\mu}_i$. Let $\hat{\boldsymbol{\theta}} = [\hat{\theta}_1 \ldots \hat{\theta}_p]^{\rm T}$ be the estimator of $\boldsymbol{\theta}$, the parameter of the regional growth curve. The estimator of the at-site quantile of nonexceedance probability F is $\hat{Q}_i(F) = \hat{\mu}_i x(F;\hat{\boldsymbol{\theta}})$.

Error in the estimated quantile comes from the variability and bias of the estimator $\hat{Q}_i(F)$. The mean square error of $\hat{Q}_i(F)$ is

$$\mathrm{E}\{\hat{Q}_i(F) - Q_i(F)\}^2 = \mathrm{var}\{\hat{Q}_i(F)\} + [\mathrm{bias}\{\hat{Q}_i(F)\}]^2\,, \tag{7.1}$$

where

$$\mathrm{bias}\{\hat{Q}_i(F)\} = \mathrm{E}\{\hat{Q}_i(F) - Q_i(F)\}\,. \tag{7.2}$$

The mean square error contains terms arising from the variability and bias of $\hat{Q}_i(F)$. We can derive approximations to these terms when the quantile estimate is obtained using the regional L-moment algorithm. The derivation uses asymptotic statistical approximations and is postponed until subsection 7.2.1. The final result is that the components of the mean square error of the quantile estimate $\hat{Q}_i(F)$ are given by

$$\mathrm{var}\{\hat{Q}_i(F)\} \approx \{x(F;\boldsymbol{\theta}_0)\}^2\,\mathrm{var}(\hat{\mu}_i) + \mu_i^2\,\mathrm{var}\{x(F;\hat{\boldsymbol{\theta}})\}\,, \tag{7.3}$$

$$\begin{aligned}\mathrm{bias}\{\hat{Q}_i(F)\} \approx{}& \mu_i\{x(F;\boldsymbol{\theta}_0) - q^{\rm R}(F)\} + \mu_i\{q^{\rm R}(F) - q_i(F)\}\\ &+ \tfrac{1}{2}\mu_i \sum_j \sum_k \left.\frac{\partial^2 x(F;\boldsymbol{\theta})}{\partial\theta_j\,\partial\theta_k}\right|_{\boldsymbol{\theta}=\boldsymbol{\theta}_0} \mathrm{cov}(\hat{\theta}_j, \hat{\theta}_k)\,.\end{aligned} \tag{7.4}$$

Expressions (7.1), (7.3), and (7.4) exhibit the quantile estimate as having variability, arising from

- variability of the sample mean, $\mathrm{var}(\hat{\mu}_i)$; and

- variability of the estimated regional growth curve, $\operatorname{var}\{x(F;\hat{\boldsymbol{\theta}})\}$, which, as shown in the derivation of Eq. (7.3), itself arises from variability of the sample L-moment ratios;

and bias, arising from

- misspecification of the regional frequency distribution, which leads to the bias term $x(F;\boldsymbol{\theta}_0) - q^{\mathrm{R}}(F)$;
- heterogeneity in the region, which leads to the bias term $q^{\mathrm{R}}(F) - q_i(F)$; and
- variability of the sample L-moment ratios, the $\operatorname{cov}(\hat{\theta}_j, \hat{\theta}_k)$ terms, which induces bias in the quantile estimate in consequence of the nonlinearity of the regional growth curve as a function of the L-moment ratios.

The expressions ignore effects arising from

- bias in the sample mean,
- bias in the sample L-moment ratios, and
- covariance between the sample mean and the estimated regional growth curve,

because these are usually negligible in practice – though some cases in which the last of these components can be significant are noted in Section 7.5.3.

The distinction that we have made between bias due to misspecification and bias due to heterogeneity is somewhat arbitrary – one might argue that any heterogeneous region is misspecified, because no single frequency distribution is appropriate for every site – and depends on the particular definition of the average growth curve $q^{\mathrm{R}}(F)$. In a heterogeneous region the aim of regional frequency analysis must be to estimate a single regional frequency distribution that does not consistently over- or under-estimate the quantiles at every site. We therefore think it useful to distinguish bias due to misspecification, which is the same at each site in the region, from bias due to heterogeneity, which varies from site to site in such a way that its average over all the sites in the region is zero. In practice it is often useful to work with the relative bias of the estimated growth curve, $\mathrm{E}\{x(F;\hat{\boldsymbol{\theta}}) - q_i(F)\}/q_i(F)$. It is then reasonable to say that the best average growth curve for a heterogeneous region should satisfy the condition that the relative bias of $q^{\mathrm{R}}(F)$ as an estimator of $q_i(F)$, averaged over all the sites in the region, should be zero, that is, that $\sum_i \{q^{\mathrm{R}}(F) - q_i(F)\}/q_i(F) = 0$. This criterion implies that the average growth curve should be defined by $q^{\mathrm{R}}(F) = N/\sum\{1/q_i(F)\}$, the harmonic mean of the at-site growth curves.

An expression related to Eqs. (7.3) and (7.4), but less detailed, has been given by Stedinger et al. (1992, Eq. (18.5.3)). More detailed analytical expressions for the bias and variance of L-moment estimators of the generalized extreme-value distribution have been given by Lu and Stedinger (1992a,b). These can be used to obtain good approximations to the variance of quantile estimates for at-site and regional frequency analyses when the true and fitted frequency distributions are both of the generalized extreme-value form (Stedinger and Lu, 1995). Corresponding results have not been obtained for other distributions, however, so for arbitrary combinations of true and fitted distributions, Monte Carlo simulation is our preferred method of establishing the properties of the regional *L*-moment algorithm.

7.2.1 Derivation of Eqs. (7.3) and (7.4)

To a first-order asymptotic approximation, assuming n_i and the total number of observations $n_R = \sum_j n_j$ to be large, the variance of $\hat{Q}_i(F)$ is given by

$$\begin{aligned}\operatorname{var}\{\hat{Q}_i(F)\} &= \operatorname{var}\{\hat{\mu}_i x(F;\hat{\boldsymbol{\theta}})\} \\ &\approx \{x(F;\boldsymbol{\theta}_0)\}^2 \operatorname{var}(\hat{\mu}_i) + 2\mu_i x(F;\boldsymbol{\theta}_0) \operatorname{cov}\{\hat{\mu}_i, x(F;\hat{\boldsymbol{\theta}})\} \\ &\quad + \mu_i^2 \operatorname{var}\{x(F;\hat{\boldsymbol{\theta}})\}. \end{aligned} \qquad (7.5)$$

Of the three terms on the right side of Eq. (7.5), the first arises from the variability of the estimator of the index flood. The second term arises from the covariance between the estimators of the index flood and the regional growth curve and is generally small; because $\hat{\mu}_i$ is calculated from only the data for site i whereas $\hat{\boldsymbol{\theta}}$ uses data from all N sites, it is reasonable to expect the correlation between them to be of order $(n_i/n_R)^{1/2}$. When the n_i at each site are approximately equal, this correlation is of order $1/\sqrt{N}$. The third term on the right side of Eq. (7.5) comes from the variability of the estimated regional growth curve. It can be further approximated as

$$\operatorname{var}\{x(F;\hat{\boldsymbol{\theta}})\} \approx \sum_j \sum_k \left.\frac{\partial x(F;\boldsymbol{\theta})}{\partial \theta_j}\right|_{\boldsymbol{\theta}=\boldsymbol{\theta}_0} \left.\frac{\partial x(F;\boldsymbol{\theta})}{\partial \theta_k}\right|_{\boldsymbol{\theta}=\boldsymbol{\theta}_0} \operatorname{cov}(\hat{\theta}_j, \hat{\theta}_k), \qquad (7.6)$$

where θ_j, $j = 1, \ldots, p$, are the elements of $\boldsymbol{\theta}$. This shows how the variability in the estimated regional growth curve arises from the variability in the estimates $\hat{\theta}_j$, which in the regional *L*-moment algorithm are the regional average *L*-moment ratios.

The bias of $\hat{Q}_i(F)$ is given by

$$\begin{aligned}\mathrm{E}\{\hat{Q}_i(F) - Q_i(F)\} &= q_i(F)\, \mathrm{E}(\hat{\mu}_i - \mu_i) + \operatorname{cov}\{\hat{\mu}_i, x(F;\hat{\boldsymbol{\theta}})\} \\ &\quad + \mu_i\, \mathrm{E}\{x(F;\hat{\boldsymbol{\theta}}) - q_i(F)\}. \end{aligned} \qquad (7.7)$$

The first term in the bias arises from the bias of $\hat{\mu}_i$. This is zero for the regional L-moment algorithm, in which the index flood is the mean of the at-site frequency distribution and is estimated by the sample mean at site i. As with the variance, the second term in the bias arises from the covariance between the estimators of the index flood and the regional growth curve and is generally negligible. The third term on the right side of Eq. (7.7) comes from the bias of the estimated regional growth curve, $x(F;\hat{\boldsymbol{\theta}})$. The expectation of $x(F;\hat{\boldsymbol{\theta}})$ can be approximated by taking expectations of the first few terms of a stochastic Taylor-series expansion of $x(F;\hat{\boldsymbol{\theta}})$ about $x(F;\boldsymbol{\theta}_0)$:

$$\mathrm{E}\{x(F;\hat{\boldsymbol{\theta}})\} \approx x(F;\boldsymbol{\theta}_0) + \sum_j \left.\frac{\partial x(F;\boldsymbol{\theta})}{\partial \theta_j}\right|_{\boldsymbol{\theta}=\boldsymbol{\theta}_0} \mathrm{E}(\hat{\theta}_j - \theta_j) + \tfrac{1}{2}\sum_j \sum_k \left.\frac{\partial^2 x(F;\boldsymbol{\theta})}{\partial \theta_j \partial \theta_k}\right|_{\boldsymbol{\theta}=\boldsymbol{\theta}_0} \mathrm{cov}(\hat{\theta}_j, \hat{\theta}_k)\,. \tag{7.8}$$

We can now approximate the bias of $x(F;\hat{\boldsymbol{\theta}})$ by a sum of four components:

$$\mathrm{E}\{x(F;\hat{\boldsymbol{\theta}}) - q_i(F)\} \approx \{x(F;\boldsymbol{\theta}_0) - q^{\mathrm{R}}(F)\} + \{q^{\mathrm{R}}(F) - q_i(F)\} + \sum_j \left.\frac{\partial x(F;\boldsymbol{\theta})}{\partial \theta_j}\right|_{\boldsymbol{\theta}=\boldsymbol{\theta}_0} \mathrm{E}(\hat{\theta}_j - \theta_j) + \tfrac{1}{2}\sum_j \sum_k \left.\frac{\partial^2 x(F;\boldsymbol{\theta})}{\partial \theta_j \partial \theta_k}\right|_{\boldsymbol{\theta}=\boldsymbol{\theta}_0} \mathrm{cov}(\hat{\theta}_j, \hat{\theta}_k)\,. \tag{7.9}$$

The first term on the right side of Eq. (7.9) is due to misspecification of the regional frequency distribution. It is the error of $x(F;\boldsymbol{\theta}_0)$ as an approximation to the true regional growth curve $q^{\mathrm{R}}(F)$. The second term on the right side of Eq. (7.9) is due to heterogeneity. It is the difference between the regional and at-site growth curves. The third term arises from the bias of the estimators $\hat{\theta}_j$, which in the regional L-moment algorithm are sample L-moment ratios. These statistics have low bias (see Figure 2.7), and this term is generally negligible. The final term in Eq. (7.9) is bias that arises from the variability of the sample L-moment ratios through the nonlinear dependence of $x(F;\boldsymbol{\theta})$ on $\boldsymbol{\theta}$. This nonlinearity is most pronounced for quantiles in the extreme tail of the distribution; for quantiles in the main body of the distribution the contribution of this term is generally negligible.

7.3 Simulation of the regional *L*-moment algorithm

As noted in Section 1.2, Monte Carlo simulation is an effective tool for establishing the properties of complex statistical procedures such as the regional *L*-moment algorithm. The general procedure starts by defining a region, that is, by specifying the number of sites in the region and the frequency distributions and record lengths at each site. Many sets of data are generated from the region, and one or more estimation methods are applied to each data set. The specification of an estimation method consists of whether the method involves regional or at-site estimation and which distribution is fitted by the method. The estimated quantiles and growth curve are compared with the true values implied by the frequency distributions specified for each site, and accuracy measures are calculated for the estimators.

The procedure for simulating the regional *L*-moment algorithm is given in Table 6.1. In the course of the simulations, quantile estimates are calculated for various nonexceedance probabilities. At the mth repetition, let the estimated regional growth curve and site-i quantile estimate for nonexceedance probability F be $\hat{q}^{[m]}(F)$ and $\hat{Q}_i^{[m]}(F)$, respectively. Then at site i the relative error of the estimated regional growth curve as an estimator of the at-site growth curve $q_i(F)$ is $\{\hat{q}^{[m]}(F) - q_i(F)\}/q_i(F)$, and the relative error of the quantile estimate for nonexceedance probability F is $\{\hat{Q}_i^{[m]}(F) - Q_i(F)\}/Q_i(F)$. These quantities can be averaged over all M repetitions to approximate the bias and RMSE of the estimators. For the comparisons that we make in this chapter, the numbers are small and it is convenient to express them as percentages. The relative bias and relative RMSE, expressed as percentages, of the site-i quantile estimator are approximated, for large M, by

$$B_i(F) = M^{-1} \sum_{m=1}^{M} \frac{\hat{Q}_i^{[m]}(F) - Q_i(F)}{Q_i(F)} \times 100\,\% \tag{7.10}$$

and

$$R_i(F) = \left[M^{-1} \sum_{m=1}^{M} \left\{ \frac{\hat{Q}_i^{[m]}(F) - Q_i(F)}{Q_i(F)} \right\}^2 \right]^{1/2} \times 100\,\%\,. \tag{7.11}$$

To obtain a summary of the performance of an estimation procedure over all of the sites in the region, we compute the regional average relative bias of the estimated quantile,

$$B^{\mathrm{R}}(F) = N^{-1} \sum_{i=1}^{N} B_i(F)\,, \tag{7.12}$$

the regional average absolute relative bias of the estimated quantile,

$$A^{\mathrm{R}}(F) = N^{-1} \sum_{i=1}^{N} |B_i(F)| , \tag{7.13}$$

and the regional average relative RMSE of the estimated quantile,

$$R^{\mathrm{R}}(F) = N^{-1} \sum_{i=1}^{N} R_i(F) . \tag{7.14}$$

The regional average relative bias, $B^{\mathrm{R}}(F)$, measures the tendency of quantile estimates to be uniformly too high or too low across the whole region. This tendency is apparent, for example, when a distribution with a heavy upper tail is fitted to a region in which the true frequency distributions have relatively light upper tails, or vice versa.

The regional average absolute relative bias, $A^{\mathrm{R}}(F)$, measures the tendency of quantile estimates to be consistently high at some sites and low at others. This occurs in a heterogeneous region, in which the estimated regional growth curve tends to overestimate the true at-site growth curve at some sites and to underestimate it at others. In such cases $A^{\mathrm{R}}(F)$ indicates the magnitude of the bias at a typical site and is more useful than $B^{\mathrm{R}}(F)$, in which the contributions of negative and positive biases may cancel out to give a misleadingly small value of the bias. In a homogeneous region, however, we would expect the bias to be the same at each site, and therefore $A^{\mathrm{R}}(F)$ and $B^{\mathrm{R}}(F)$ to be equal.

The regional average relative RMSE, $R^{\mathrm{R}}(F)$, measures the overall deviation of estimated quantiles from true quantiles. It is the criterion to which we give most weight in judging whether one estimation procedure is superior to another.

In addition to the overall accuracy measures of quantile estimates, Eqs. (7.12)–(7.14), we also calculate corresponding quantities for each site's growth curve estimate. These quantities are defined by Eqs. (7.10)–(7.14) but with $Q_i(F)$ and $\hat{Q}_i^{[m]}(F)$ replaced by $q_i(F)$ and $\hat{q}^{[m]}(F)$, respectively. Comparison of the accuracy of the estimated growth curve and the estimated quantiles facilitates judgement of the relative importance of errors in estimating the index flood and errors in estimating the regional growth curve. Accuracy measures for the growth curve are also relevant when only the growth curve estimate is of interest, as noted in Section 6.4.

The number of repetitions, M, of the simulation procedure must be large enough that the bias and RMSE measures $B_i(F)$ and $R_i(F)$ are close to the true bias and RMSE, $\mathrm{E}[\{\hat{Q}_i(F) - Q_i(F)\}/Q_i(F)]$ and $(\mathrm{E}[\{\hat{Q}_i(F) - Q_i(F)\}/Q_i(F)]^2)^{1/2}$, respectively. This enables reliable comparisons to be made between the performance measures (7.10)–(7.14) for different regions. We use $M = 10{,}000$, which for the representative regions defined in Section 7.4, generally gives an accuracy of 1 or 2 units in the third decimal place for relative RMSE measures and 3 or 4 units in the third decimal place for relative bias measures; accuracy is less when the relative RMSE is large, 50% or greater.

In the simulation results presented in the tables and figures in Sections 7.4 and 7.5, bias and RMSE values are shown as percentages of the true value of the quantile or growth curve, as noted above. In the figures, RMSE values are plotted on a logarithmic scale because it is the proportional difference between RMSEs that yields the most informative comparison of different estimators.

7.4 Simulation results for representative regions

We use four regions as representative examples for illustrating the performance of the regional L-moment algorithm. Each region has 15 sites with record length 30 at each site. The frequency distributions at the sites are chosen to be representative of typical applications of regional frequency analysis.

The first two regions are plausible models for moderately to highly skew data on environmental extremes such as annual maximum instantaneous streamflow or rainfall. The frequency distribution at each site is a generalized extreme-value (GEV) distribution. The first region, Region R1, is homogeneous, with each site having L-CV 0.25 and L-skewness 0.25, corresponding to a GEV shape parameter $k = -0.121$. The second region, Region R2, is similar to the first but heterogeneous, with L-CV and L-skewness both varying linearly from 0.20 at site 1 to 0.30 at site 15. This form of heterogeneity is used because it should be realistic. Regions will often not be distinct entities but will be constructed from sites whose frequency distribution span a continuum of shapes. Linear variation of L-moment ratios is then a plausible form of variation for the sites in a region. In practical applications it will often be the case that sites with high L-CV also tend to have high L-skewness. For example, Lu and Stedinger (1992b) found this to be the case for annual maximum streamflow data.

Simulated samples from Region R2 yield, on average, a value of $H = 1.74$ for the heterogeneity measure (4.5), and 0.54 and 0.24 for the corresponding measure based on V_2 and V_3 defined in Eqs. (4.6) and (4.7). This amount of heterogeneity could easily arise from the methods of forming regions described in Chapter 4; the values of the heterogeneity measures are typical of those encountered in practice for regions that appear to be on the borderline of heterogeneity but that cannot be divided into physically and geographically meaningful subregions that are acceptably homogeneous. The L-moment ratios, distribution parameters, and some quantiles for Region R2 are given in Table 7.1. The quantile functions for sites 1, 8, and 15

Table 7.1. *Specification of Region R2.*

				GEV parameters			Quantiles				
Site	τ	τ_3	n_i	ξ	α	k	0.01	0.1	0.9	0.99	0.999
1	0.2000	0.2000	30	0.828	0.276	–0.046	0.421	0.602	1.482	2.242	3.073
2	0.2071	0.2071	30	0.820	0.283	–0.057	0.407	0.590	1.499	2.307	3.214
3	0.2143	0.2143	30	0.812	0.289	–0.068	0.393	0.578	1.515	2.374	3.361
4	0.2214	0.2214	30	0.805	0.296	–0.079	0.379	0.566	1.532	2.442	3.515
5	0.2286	0.2286	30	0.797	0.302	–0.089	0.366	0.555	1.548	2.512	3.677
6	0.2357	0.2357	30	0.789	0.307	–0.100	0.354	0.543	1.565	2.584	3.846
7	0.2429	0.2429	30	0.781	0.313	–0.110	0.341	0.532	1.581	2.657	4.024
8	0.2500	0.2500	30	0.773	0.318	–0.121	0.329	0.521	1.597	2.732	4.210
9	0.2571	0.2571	30	0.765	0.324	–0.131	0.318	0.510	1.612	2.809	4.404
10	0.2643	0.2643	30	0.757	0.329	–0.142	0.306	0.499	1.628	2.888	4.608
11	0.2714	0.2714	30	0.749	0.333	–0.152	0.295	0.488	1.643	2.968	4.821
12	0.2786	0.2786	30	0.741	0.338	–0.162	0.284	0.478	1.658	3.050	5.044
13	0.2857	0.2857	30	0.733	0.342	–0.173	0.274	0.467	1.673	3.134	5.277
14	0.2929	0.2929	30	0.725	0.346	–0.183	0.264	0.457	1.688	3.220	5.521
15	0.3000	0.3000	30	0.717	0.350	–0.193	0.254	0.447	1.702	3.307	5.775

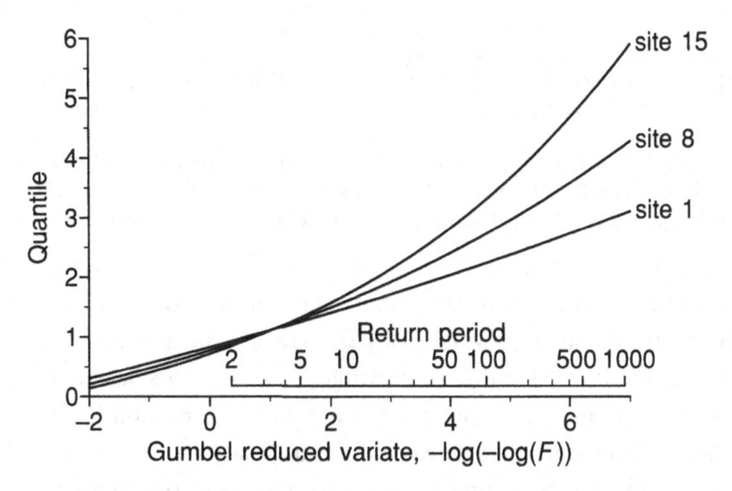

Fig. 7.1. Quantile functions for three sites in Region R2.

are illustrated in Figure 7.1. In Region R1, of course, each site has parameters and quantiles the same as those of site 8 in Region R2.

Sample *L*-moment ratios for the 15 sites in Regions R1 and R2 are shown in Figure 7.2 for a typical realization of these regions. Sampling variability causes these sample *L*-moment ratios to be much more widely scattered than the population *L*-moment ratios. The greater spread in the sample *L*-CV values for the heterogeneous Region R2 than in the homogeneous Region R1 is apparent, but the spread

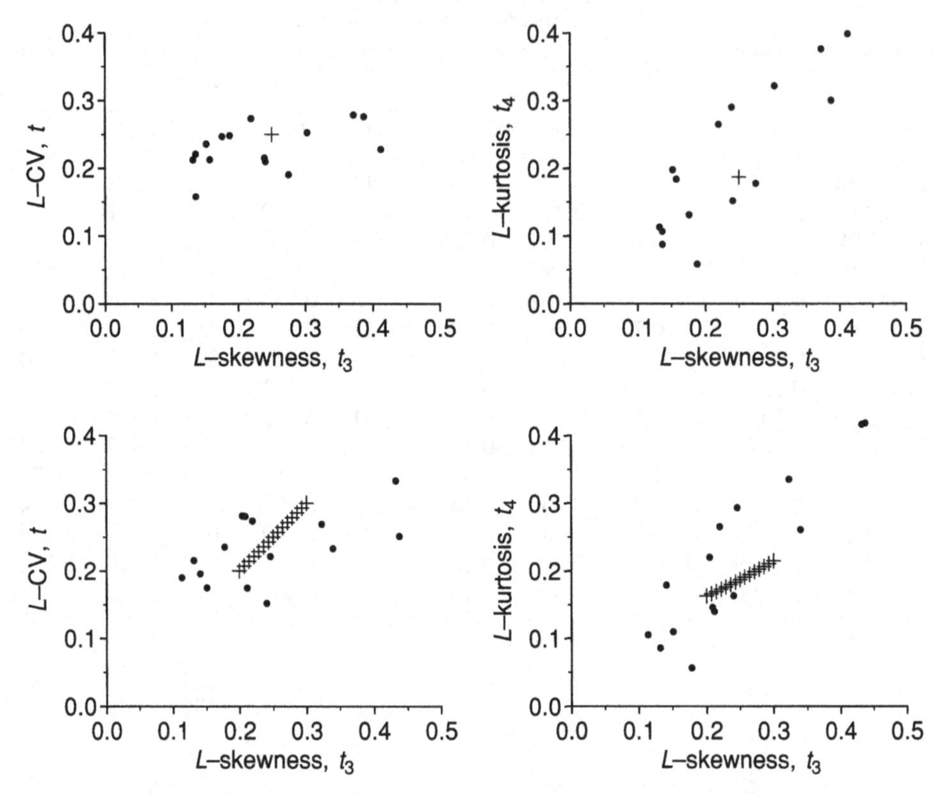

Fig. 7.2. Sample L-moment ratios for typical realizations of Region R1 (upper graphs) and Region R2 (lower graphs). Population L-moment ratios for the sites are also shown (+ marks). For Region R1, all sites have the same population L-moment ratios.

of the L-skewness and L-kurtosis values is very similar for the two regions. This is consistent with the average values of heterogeneity measures described earlier. It also explains the limited ability of heterogeneity measures based on L-skewness and L-kurtosis to identify heterogeneity in a region like Region R2, for which the between-site variations in L-skewness and L-kurtosis are small compared with the sampling variability of the sample L-skewness and L-kurtosis statistics.

A regional generalized extreme-value distribution was fitted, using the regional L-moment algorithm, to data simulated from Regions R1 and R2. Simulation results are summarized in Table 7.2 and Figure 7.3. Several features characteristic of the performance of the regional L-moment algorithm are exhibited by these results. We consider first the results for the homogeneous Region R1. Quantile estimates become less and less accurate at larger return periods. There is little bias in the quantile estimates; because the region is homogeneous and correctly specified, the only significant source of bias is "nonlinearity bias," the last term in Eq. (7.4). In addition to this bias, the RMSE of the estimated growth curve contains a contribution

Table 7.2. *Simulation results for Regions R1 and R2.*

		Quantiles			Growth curve		
Region	F:	0.9	0.99	0.999	0.9	0.99	0.999
	Bias	−0.2	−2.0	−3.9	−0.2	−2.0	−3.9
R1	Abs. bias	0.2	2.0	3.9	0.2	2.0	3.9
	RMSE	9.2	11.0	14.6	1.3	5.7	11.0
	Bias	0.1	−1.2	−2.1	0.1	−1.2	−2.2
R2	Abs. bias	3.7	10.4	16.6	3.7	10.4	16.6
	RMSE	10.2	15.7	23.0	4.1	12.2	20.6

Note: Tabulated values are the regional average relative bias, absolute relative bias, and relative RMSE of estimated quantiles, expressed as percentages, that is, $B^{\mathrm{R}}(F)$, $A^{\mathrm{R}}(F)$, and $R^{\mathrm{R}}(F)$ as defined in Eqs. (7.12)–(7.14), and the corresponding quantities for the estimated growth curve.

from the variability of the estimated growth curve, and the RMSE of estimated quantiles contains a further contribution from the variability of the estimated index flood (the at-site sample mean). Variability of the mean is the most important factor for quantiles in the main body of the distribution: for example, at $F = 0.9$ in Table 7.2, including the variability of the mean raises the regional average relative RMSE from 1.3% to 9.2%. Far into the tail of the distribution, variability of the estimated growth curve is more important. At $F = 0.999$ in Table 7.2, including the variability of the mean now raises the regional average relative RMSE only from 11.0% to 14.6%.

Turning to the results for Region R2, we see that the main effect of heterogeneity is to introduce bias into the estimated growth curve, which also adds bias to the estimated quantiles. This bias is positive at sites where the true quantile is less than the average for the region and negative where the true quantile is greater than the regional average. At the extreme sites, 1 and 15, this bias is the dominant contribution to the RMSE of the estimated quantiles. Averaged over the region as a whole, its effect is large but not overwhelming; the regional average relative RMSEs of estimated quantiles are larger for Region R2 than for Region R1 by 11% at $F = 0.9$, 40% at $F = 0.99$, and 50% at $F = 0.999$.

Because it is difficult in practice to ensure that the sites used in a particular application of regional frequency analysis constitute a region that is exactly homogeneous, the utility of regional frequency analysis depends on whether its performance is acceptable for moderately heterogeneous regions such as Region R2. The performance of the regional *L*-moment algorithm should be acceptable for many applications. In Region R2, for example, the regional *L*-moment algorithm

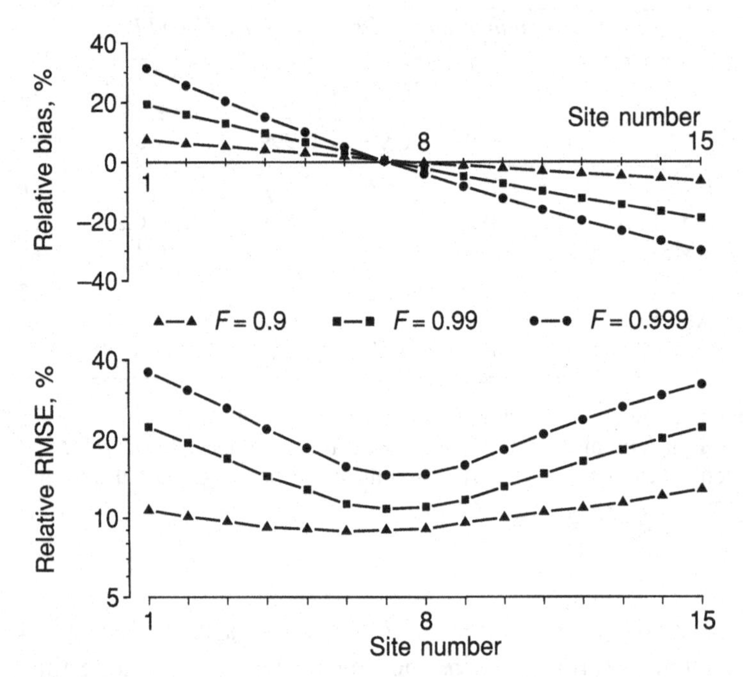

Fig. 7.3. Simulation results for the sites in Region R2. Plotted values are the relative bias and relative RMSE of the quantile estimator $\hat{Q}_i(F)$, that is, $B_i(F)$ and $R_i(F)$ as defined in Eqs. (7.10) and (7.11), for the 15 sites in Region R2.

estimates the $F = 0.99$ quantile (for annual data, the 100-year event) with RMSE 11% to 22% of the true value of the quantile. In particular – see Section 7.5.3 – the regional *L*-moment algorithm is certainly superior to estimation methods that use only the at-site data.

Our third and fourth example regions are plausible models for data, such as annual precipitation totals, that are positive but only slightly skew. The frequency distribution at each site is a lognormal distribution. Again, the regions form a homogeneous–heterogeneous pair. Region R3 is homogeneous, with each site having *L*-CV $\tau = 0.08$ and *L*-skewness $\tau_3 = 0.05$, corresponding to a lognormal shape parameter $k = -0.102$ and conventional skewness $\gamma = 0.31$. Region R4 is heterogeneous, with *L*-CV varying linearly from 0.065 at site 1 to 0.095 at site 15 and *L*-skewness varying linearly from 0 at site 1 to 0.1 at site 15. Simulated samples from this region yield, on average, a value of $H = 1.81$ for the heterogeneity measure (4.5). Again, this amount of heterogeneity could easily arise in regions formed by the methods described in Chapter 4. The *L*-moment ratios, distribution parameters, and some quantiles for Region R4 are given in Table 7.3. The quantile functions for sites 1, 8, and 15 are illustrated in Figure 7.4. In Region R3, each site has parameters and quantiles the same as those of site 8 in Region R4.

Table 7.3. *Specification of Region R4.*

				Lognormal parameters			Quantiles				
Site	τ	τ_3	n_i	ξ	α	k	0.01	0.1	0.9	0.99	0.999
1	0.0650	0.0000	30	1.000	0.115	0.000	0.732	0.852	1.148	1.268	1.356
2	0.0671	0.0071	30	0.999	0.119	–0.015	0.727	0.848	1.153	1.281	1.375
3	0.0693	0.0143	30	0.998	0.123	–0.029	0.722	0.844	1.159	1.294	1.395
4	0.0714	0.0214	30	0.997	0.127	–0.044	0.717	0.840	1.164	1.307	1.416
5	0.0736	0.0286	30	0.996	0.130	–0.059	0.713	0.835	1.169	1.321	1.437
6	0.0757	0.0357	30	0.995	0.134	–0.073	0.709	0.831	1.175	1.335	1.459
7	0.0779	0.0429	30	0.994	0.138	–0.088	0.704	0.827	1.181	1.349	1.482
8	0.0800	0.0500	30	0.993	0.141	–0.102	0.701	0.823	1.186	1.364	1.506
9	0.0821	0.0571	30	0.992	0.145	–0.117	0.697	0.819	1.192	1.379	1.530
10	0.0843	0.0643	30	0.990	0.148	–0.132	0.693	0.815	1.197	1.394	1.556
11	0.0864	0.0714	30	0.989	0.152	–0.146	0.689	0.811	1.203	1.410	1.582
12	0.0886	0.0786	30	0.987	0.155	–0.161	0.686	0.808	1.208	1.426	1.609
13	0.0907	0.0857	30	0.986	0.159	–0.176	0.683	0.804	1.214	1.442	1.637
14	0.0929	0.0929	30	0.984	0.162	–0.190	0.680	0.800	1.220	1.459	1.667
15	0.0950	0.1000	30	0.983	0.166	–0.205	0.677	0.796	1.225	1.476	1.697

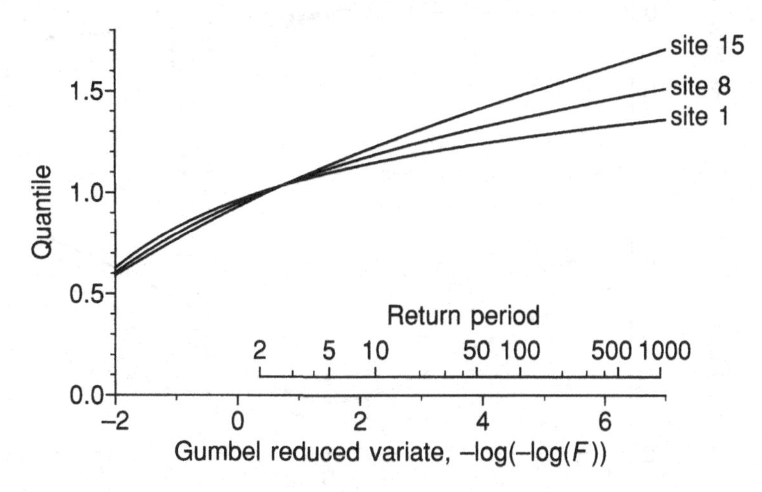

Fig. 7.4. Quantile functions for three sites in Region R4.

A regional lognormal distribution was fitted, using the regional *L*-moment algorithm, to data simulated from Regions R3 and R4. Simulation results are summarized in Table 7.4 and Figure 7.5. Note that the vertical scales in Figures 7.3 and 7.5 are different.

The results are similar in several respects to those for Regions R1 and R2. Estimators have rapidly decreasing accuracy as the return period increases. The relative contributions of errors in estimating the index flood and errors in estimating

Table 7.4. *Simulation results for Regions R3 and R4.*

		Quantiles			Growth curve		
Region	F:	0.9	0.99	0.999	0.9	0.99	0.999
	Bias	0.0	−0.1	−0.1	0.0	−0.1	−0.1
R3	Abs. bias	0.0	0.1	0.1	0.0	0.1	0.1
	RMSE	2.7	3.0	3.7	0.6	1.6	2.6
	Bias	0.0	−0.1	−0.1	0.0	−0.1	−0.1
R4	Abs. bias	1.8	4.1	6.0	1.8	4.1	6.0
	RMSE	3.3	5.3	7.3	1.9	4.5	6.7

Note: Tabulated values are the regional average relative bias, absolute relative bias, and relative RMSE of estimated quantiles, that is, $B^R(F)$, $A^R(F)$, and $R^R(F)$ as defined in Eqs. (7.12)–(7.14), and the corresponding quantities for the estimated growth curve.

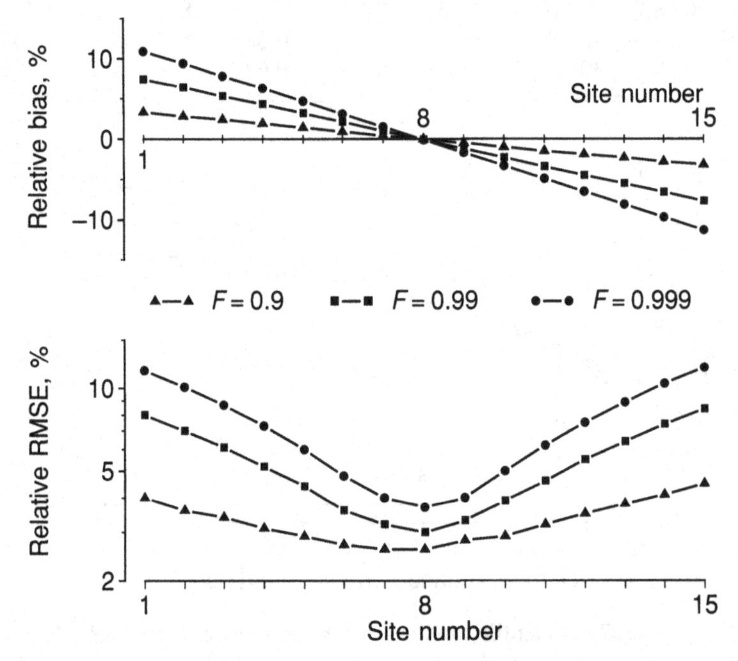

Fig. 7.5. Simulation results for the sites in Region R4. Plotted values are the relative bias and relative RMSE of the quantile estimator $\hat{Q}_i(F)$, that is, $B_i(F)$ and $R_i(F)$ as defined in Eqs. (7.10) and (7.11), for the 15 sites in Region R4.

the growth curve are similar to those for Regions R1 and R2. Bias in the estimated growth curve is the major contribution to RMSE of quantile estimates in the heterogeneous region.

Some differences can be observed, however, for Regions R3 and R4 compared with Regions R1 and R2. All relative RMSE values for Regions R3 and R4 are lower than the corresponding values for Regions R1 and R2, as extreme quantiles can be estimated with much greater accuracy for distributions with low skewness. Bias in the estimated quantiles and growth curve is entirely negligible even at extreme return periods, as nonlinearity of the quantile function is much less for distributions with low skewness. The proportion of the RMSE of estimated quantiles that can be attributed to bias in the estimated growth curve, that is, to heterogeneity, is somewhat larger for Region R4 than for Region R2, although the average heterogeneity measures for realizations of these regions are very similar.

Overall, the performance of the regional *L*-moment algorithm for Regions R3 and R4 is fairly similar to its performance for Regions R1 and R2. We can therefore expect this performance, and the way in which it is affected by the various factors considered in the following subsections, to be qualitatively similar for a wide range of region specifications and frequency distributions.

7.5 Simulation results for effects of different factors

7.5.1 Regional averaging of L-moments versus L-moment ratios

In Section 6.2 it was noted that some early versions of the index-flood procedure based on probability weighted moments calculated regional averages of the at-site *L*-moments using Eq. (6.6) rather than regional averages of the at-site *L*-moment ratios using Eqs. (6.2) and (6.3). Table 7.5 compares the performance of these two variants of the regional *L*-moment algorithm for quantile estimation in the four representative regions defined in Section 7.4. The tabulated values are the regionally averaged performance measures (7.12)–(7.14). Differences between the two variants are small, but the RMSE values for the "average *L*-moment ratios" variant are never larger than the corresponding values for the "average *L*-moments" variant. We therefore prefer to use the average *L*-moment ratios variant in all applications of regional frequency analysis.

7.5.2 "Unbiased" versus plotting-position estimators

The *L*-moment ratios used in the regional *L*-moment algorithm can be computed using the plotting-position estimators described in Section 2.8 rather than the "unbiased" estimators of at-site *L*-moment ratios, $t^{(i)}$ and $t_r^{(i)}$. Table 7.6 compares the performance of these two variants of the regional *L*-moment algorithm for quantile estimation in the four representative regions defined in Section 7.4. The plotting position used was $p_{j:n} = (j - 0.35)/n$, as recommended for the generalized extreme-value distribution by Hosking et al. (1985b). The values tabulated in Table 7.6 are the regionally averaged performance measures (7.12)–(7.14).

Table 7.5. *Comparison of variants of the regional L-moment algorithm involving regional averaging of L-moments versus L-moment ratios.*

Region		F:	Average *L*-moment ratios			Average *L*-moments		
			0.9	0.99	0.999	0.9	0.99	0.999
	Bias		−0.2	−2.0	−3.9	−0.3	−0.8	−0.9
R1	Abs. bias		0.2	2.0	3.9	0.3	0.8	0.9
	RMSE		9.2	11.0	14.6	9.2	11.0	14.9
	Bias		0.1	−1.2	−2.1	0.0	0.5	2.1
R2	Abs. bias		3.7	10.4	16.6	3.7	10.4	17.0
	RMSE		10.2	15.7	23.0	10.2	16.0	24.2
	Bias		0.0	−0.1	−0.1	0.0	0.0	0.0
R3	Abs. bias		0.0	0.1	0.1	0.0	0.0	0.0
	RMSE		2.7	3.0	3.7	2.7	3.0	3.7
	Bias		0.0	−0.1	−0.1	0.0	0.2	0.4
R4	Abs. bias		1.8	4.1	6.0	1.8	4.1	6.0
	RMSE		3.3	5.3	7.3	3.3	5.3	7.4

Note: Tabulated values are the regional average relative bias, absolute relative bias, and relative RMSE of estimated quantiles, that is, $B^{R}(F)$, $A^{R}(F)$, and $R^{R}(F)$ as defined in Eqs. (7.12)–(7.14).

In Regions R1 and R2 the plotting-position estimators perform slightly better than the "unbiased" estimators. In Regions R3 and R4 they perform considerably worse, largely on account of their higher bias. (In technical terms, this bias arises from the third term on the right side of Eq. (7.9), which is not negligible when $\hat{\theta}_j$ is a plotting-position estimator.) These simulation results indicate no clear preference between "unbiased" and plotting-position estimators. As noted in Section 2.8, "unbiased" estimators have less bias and are therefore superior for summarizing data samples and computing the heterogeneity and goodness-of-fit measures defined in Chapters 4 and 5. In the interests of consistency we prefer to use the "unbiased" estimators for fitting distributions too.

7.5.3 Regional versus at-site estimation

A fundamental comparison for assessing the utility of regional frequency analysis is that between at-site and regional estimation. The contributions of the major sources of error in quantile estimation are summarized in Figure 7.6. Because data from several or many sites are available for estimation of the regional growth curve, its variability is often much lower than the variability of the growth curve estimated from at-site data. In a heterogeneous region, however, the regional growth curve is

Table 7.6. *Comparison of variants of the regional L-moment algorithm involving "unbiased" and plotting-position estimators.*

Region		*F*:	"Unbiased"			PP		
			0.9	0.99	0.999	0.9	0.99	0.999
	Bias		−0.2	−2.0	−3.9	0.1	−1.2	−2.6
R1	Abs. bias		0.2	2.0	3.9	0.1	1.2	2.6
	RMSE		9.2	11.0	14.6	9.2	10.7	13.9
	Bias		0.1	−1.2	−2.1	0.3	−0.4	−0.9
R2	Abs. bias		3.7	10.4	16.6	3.7	10.4	16.7
	RMSE		10.2	15.7	23.0	10.2	15.6	22.7
	Bias		0.0	−0.1	−0.1	1.5	3.1	4.4
R3	Abs. bias		0.0	0.1	0.1	1.5	3.1	4.4
	RMSE		2.7	3.0	3.7	3.1	4.4	5.8
	Bias		0.0	−0.1	−0.1	1.5	3.1	4.4
R4	Abs. bias		1.8	4.1	6.0	2.1	4.8	7.0
	RMSE		3.3	5.3	7.3	3.7	6.0	8.3

Note: Tabulated values are $B^{R}(F)$, $A^{R}(F)$, and $R^{R}(F)$ as defined in Eqs. (7.12)–(7.14). Unbiased denotes unbiased estimators; PP denotes plotting-position estimators.

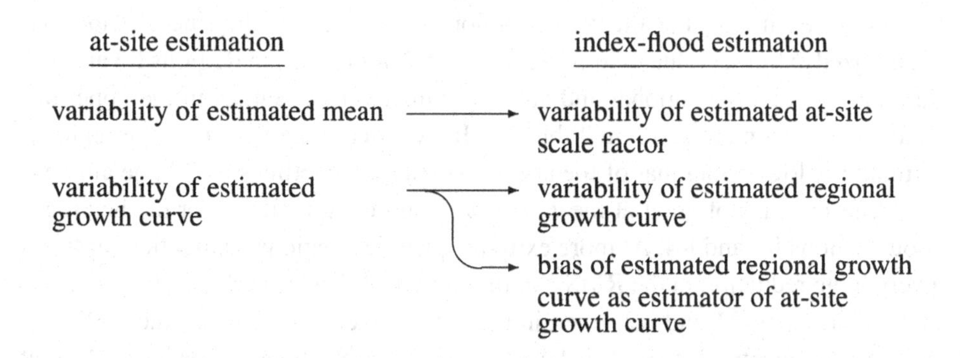

Fig. 7.6. Sources of error for at-site and index-flood estimation methods.

biased as an estimator of the at-site growth curve. This is an additional source of error in quantile estimates, especially for sites that are not typical of the region as a whole. For regional frequency analysis to be worthwhile the bias due to heterogeneity must not be so large as to outweigh the reduction in variability of the estimated growth curve.

Our simulation results compare at-site estimation using the method of *L*-moments with regional estimation using the regional *L*-moment algorithm, for Regions

Table 7.7. *Comparison of regional and at-site estimators.*

		Regional			At-site		
Region	F:	0.9	0.99	0.999	0.9	0.99	0.999
	Bias	–0.2	–2.0	–3.9	–0.8	0.6	7.8
R1	Abs. bias	0.2	2.0	3.9	0.8	0.6	7.8
	RMSE	9.2	11.0	14.6	11.8	26.9	56.8
	Bias	0.1	–1.2	–2.1	–0.8	0.6	7.7
R2	Abs. bias	3.7	10.4	16.6	0.8	0.6	7.7
	RMSE	10.2	15.7	23.0	11.8	27.0	57.2
	Bias	0.0	–0.1	–0.1	–0.1	0.4	1.2
R3	Abs. bias	0.0	0.1	0.1	0.1	0.4	1.2
	RMSE	2.7	3.0	3.7	3.5	6.7	11.0
	Bias	0.0	–0.1	–0.1	–0.1	0.4	1.2
R4	Abs. bias	1.8	4.1	6.0	0.1	0.4	1.2
	RMSE	3.3	5.3	7.3	3.5	6.8	11.2

Note: Tabulated values are the regional average relative bias, absolute relative bias and relative RMSE of estimated quantiles, that is, $B^{R}(F)$, $A^{R}(F)$, and $R^{R}(F)$ as defined in Eqs. (7.12)–(7.14).

R1–R4. In each case the fitted distribution is the same as the true distribution: generalized extreme-value in Regions R1 and R2, lognormal in Regions R3 and R4. Summary results for regional and at-site estimation of quantiles in the upper tail of the distribution are given in Table 7.7. In every case the RMSE of the regional estimator is lower than that of the at-site estimator, sometimes by a large amount. The difference is not great at the $F = 0.9$ quantile, particularly for the heterogeneous regions R2 and R4. At more extreme quantiles, regional estimation provides a very large reduction in the RMSE of the estimates. This is particularly true in the high-CV regions R1 and R2. Even in the heterogeneous region R2, the RMSE of the regional quantile estimator is lower than that of the at-site estimator by 42% at the $F = 0.99$ quantile and by 60% at the $F = 0.999$ quantile.

A more detailed picture of the utility of regional frequency analysis can be obtained from simulation results for different quantiles and for the individual sites in a heterogeneous region. A selection of the results is shown in Figures 7.7–7.10. In each of these diagrams, the vertical scale for RMSE is logarithmic and the horizontal axis is plotted as though on extreme-value probability paper, that is, it is linear in $-\log(-\log F)$, as in Figures 7.1 and 7.4.

Figure 7.7 shows the bias and RMSE of estimated quantiles for Region R1. The same graphs apply to every site in the region, because the region is homogeneous. The advantage of regional estimation for extreme quantiles, $F < 0.1$ and $F > 0.9$, is clearly apparent. For some quantiles in the main body of the distribution, $0.1 \le F \le 0.7$, at-site estimation is more accurate than regional estimation even in this homogeneous region. This occurs in regions in which the frequency distribution is fairly skew, and is a consequence of interaction between the estimates of the index flood and the growth curve; the covariance in the second term on the right side of Eq. (7.3) is negative and is larger for at-site than for regional estimators of the distribution parameters.

Figure 7.8 shows corresponding results for three sites in Region R2; site 8 has *L*-moment ratios that are the same as the regional average, whereas the extreme sites, 1 and 15, differ the most from the average. Site 8 benefits most from regional estimation; its RMSE curve is even a little lower than that for Region R1. At the extreme sites, particularly site 1, regional estimation gives lower RMSEs only for the more extreme quantiles, for which at-site estimators have high variability.

Figure 7.9 is analogous to Figure 7.7 but shows results for Region R3. The general pattern is similar to that of Region R1, except that regional estimation is advantageous for all quantiles. The magnitudes of bias and RMSE are also much smaller for this low-CV region than for the high-CV Region R1.

Figure 7.10 is analogous to Figure 7.8 but shows results for sites 1, 8, and 15 in Region R4. The general pattern is similar to that of Region R2, except that at-site estimation for site 1 remains preferable to regional estimation out to and beyond the $F = 0.999$ quantile. Results not illustrated show that the same is true for sites 2 and 3. Thus although regional estimation may be preferable for the region as a whole, at-site estimation may still give better performance at individual sites. A practical consequence is that if one site is of particular interest, care should be taken to ensure that it is not markedly atypical of the region to which it is assigned.

7.5.4 Number of sites in region

The extent to which regional frequency analysis is preferable to at-site analysis depends on N, the number of sites in the region. Homogeneous regions with varying N were generated in which all sites have the same generalized extreme-value frequency distribution as in Region R1 and record length 30. Figure 7.11 summarizes simulation results for regional estimation, fitting a generalized extreme-value distribution with the regional *L*-moment algorithm, of the quantiles and growth curve for nonexceedance probabilities $F = 0.9$, 0.99, and 0.999. Growth curve estimation using the regional data becomes increasingly accurate as the region size increases. The rate of decrease in the RMSE is close to the theoretical asymptotic rate of $n_{\mathrm{R}}^{-1/2}$, where $n_{\mathrm{R}} = \sum_i n_i$ is the total number of data points for the region; because the record lengths are the same at each site, this decay rate is also equivalent

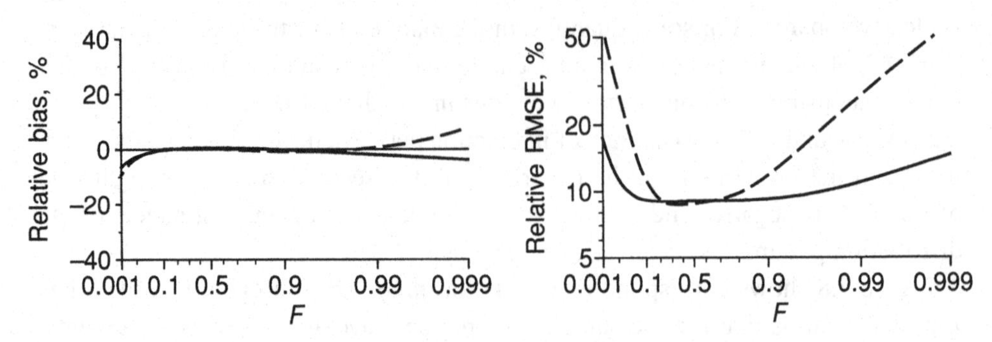

Fig. 7.7. Relative bias and RMSE of estimated quantiles for sites in Region R1. Fitted distribution: GEV. Estimation methods: at-site (dashed line) and regional (solid line).

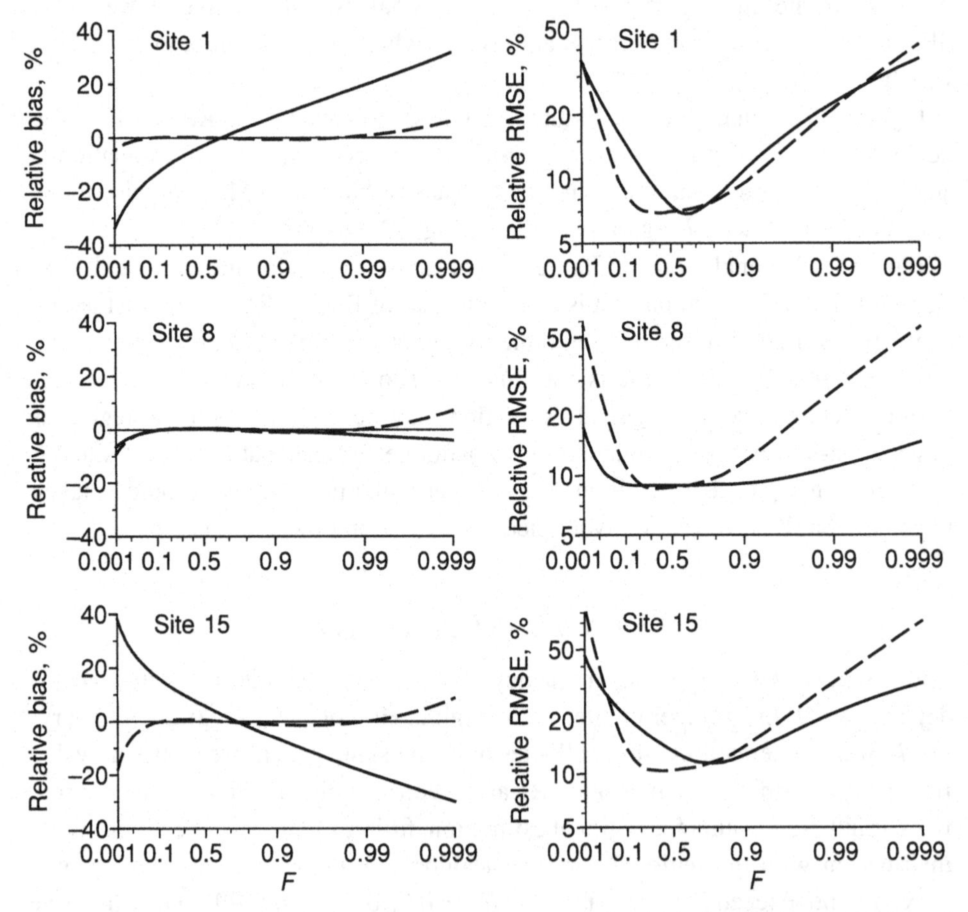

Fig. 7.8. Relative bias and RMSE of estimated quantiles for three sites in Region R2. Fitted distribution: GEV. Estimation methods: at-site (dashed line) and regional (solid line).

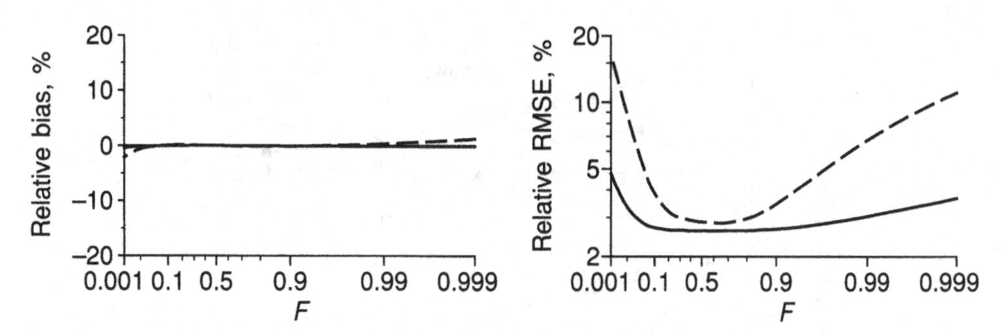

Fig. 7.9. Relative bias and RMSE of estimated quantiles for sites in Region R3. Fitted distribution: lognormal. Estimation methods: at-site (dashed line) and regional (solid line).

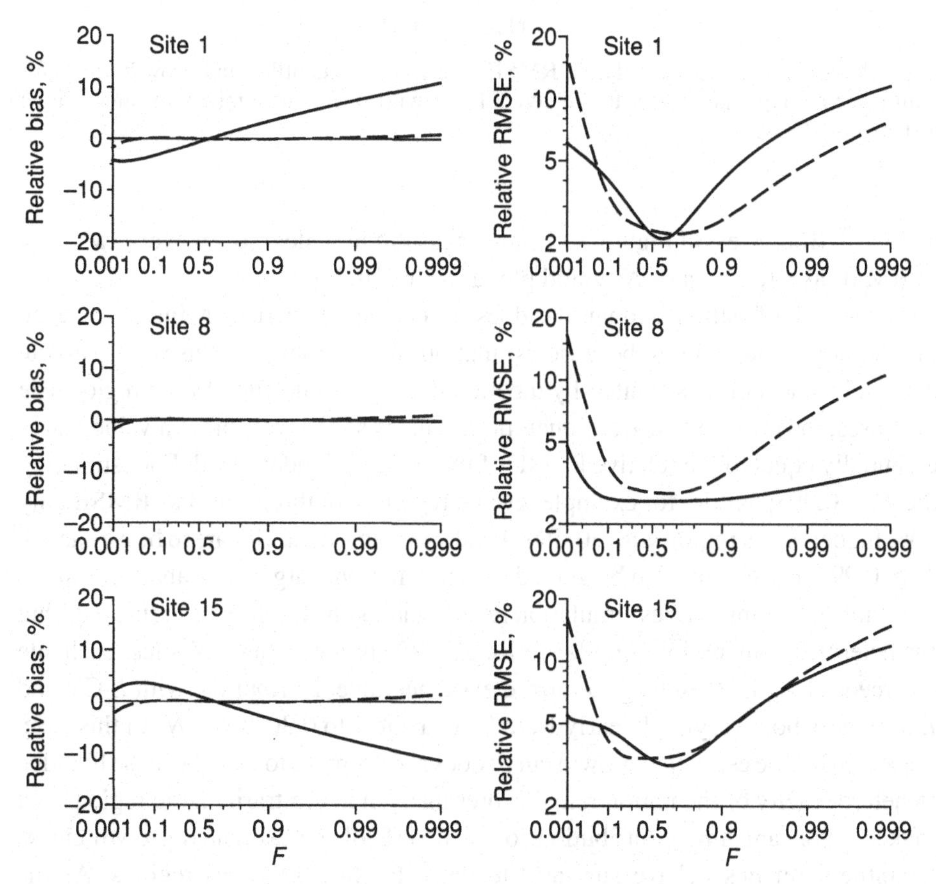

Fig. 7.10. Relative bias and RMSE of estimated quantiles for three sites in Region R4. Fitted distribution: lognormal. Estimation methods: at-site (dashed line) and regional (solid line).

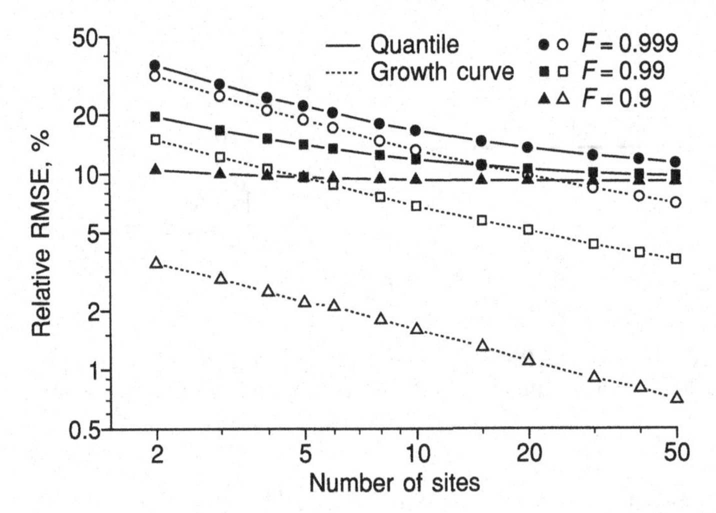

Fig. 7.11. Regional average relative RMSE of estimated quantiles and growth curve for homogeneous regions similar to Region R1 but with different numbers of sites. Fitted distribution: GEV.

to $N^{-1/2}$. Both axes on Figure 7.11 have logarithmic scales, so a power-law decay of RMSE as a function of N would plot as a straight line.

The RMSE of estimated quantiles does not decrease to zero as N increases, even in a homogeneous region, because estimation of the index flood uses only at-site data and its accuracy is limited by the record length at the site. As the region size increases, the relative RMSE of quantile estimates soon levels off at a value that is essentially equal to the relative RMSE of the estimated index flood. For estimating the $F = 0.99$ quantile, for example, even a region with three sites has RMSE only 10% higher than a region with 50 sites. Unless extreme quantiles are to be estimated, $F \geq 0.999$, there is little to be gained by using regions larger than about 20 sites.

Figure 7.12 summarizes simulation results analogous to those of Figure 7.11 but for heterogeneous regions similar to Region R2. In a region of N sites, each site has record length 30 and a generalized extreme-value distribution, with L-CV and L-skewness both varying linearly from 0.20 at site 1 to 0.30 at site N. In this case, the RMSE of the estimated growth curve does not decrease to zero, because bias due to heterogeneity of the region remains present even in arbitrarily large regions and makes a nonvanishing contribution to the RMSE of the estimated growth curve. Quantile estimates behave similarly to those for homogeneous regions. As the region size increases, the relative RMSE approaches a limit that now results from a combination of variability of the index-flood estimator and bias due to heterogeneity.

Analogous simulations were also run for regions similar to Regions R3 and R4 but with different numbers of sites. The results were qualitatively similar to those

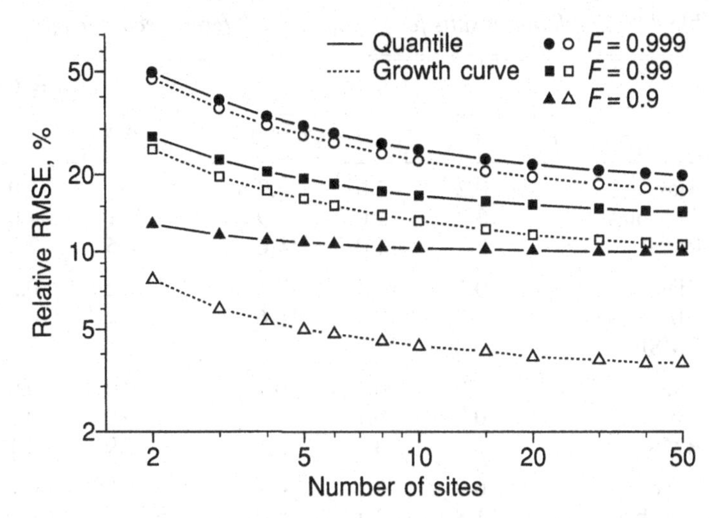

Fig. 7.12. Regional average relative RMSE of estimated quantiles and growth curve for heterogeneous regions similar to Region R2 but with different numbers of sites. Fitted distribution: GEV.

shown in Figures 7.11 and 7.12 and are not illustrated separately. The general conclusion, which remains valid for a wide range of regions, is that unless extreme quantiles are to be estimated, $F \geq 0.999$, there is little to be gained by using regions larger than about 20 sites.

7.5.5 Record length

Statistical estimators are more accurate when calculated from a large sample of data. For at-site frequency analysis, the bias and variance of estimated quantiles and growth curve are approximately proportional to n_i^{-1}, and the RMSE of these estimators is therefore approximately proportional to $n_i^{-1/2}$.

In regional frequency analysis using the regional *L*-moment algorithm, the situation is more complicated. The major components of error in quantile estimates have different dependencies on sample size. The estimated index flood at site i, based on data from that site only, has variance approximately proportional to n_i^{-1}. The estimated regional growth curve has variance approximately proportional to n_R^{-1}, where $n_R = \sum n_i$ is the total number of observations available in the region. In heterogeneous regions, the regional growth curve is biased as an estimator of the growth curve at any individual site, and this bias is independent of the sample size.

The effect on the regional *L*-moment algorithm of changing the record length at each site in a region is illustrated in Table 7.8. The table gives regional average performance measures of quantile estimators for the representative regions R1–R4,

Table 7.8. *Simulation results for regions with different record lengths.*

Region		F:	$n = 30$			$n = 60$		
			0.9	0.99	0.999	0.9	0.99	0.999
	Bias		–0.2	–2.0	–3.9	–0.1	–1.0	–2.0
R1	Abs. bias		0.2	2.0	3.9	0.1	1.0	2.0
	RMSE		9.2	11.0	14.6	6.5	7.8	10.5
	Bias		0.1	–1.2	–2.1	0.2	–0.2	–0.2
R2	Abs. bias		3.7	10.4	16.6	3.7	10.4	16.8
	RMSE		10.2	15.7	23.0	7.8	13.6	20.6
	Bias		0.0	–0.1	–0.1	0.0	0.0	–0.1
R3	Abs. bias		0.0	0.1	0.1	0.0	0.0	0.1
	RMSE		2.7	3.0	3.7	1.9	2.1	2.6
	Bias		0.0	–0.1	–0.1	0.0	–0.1	–0.1
R4	Abs. bias		1.8	4.1	6.0	1.7	4.0	6.0
	RMSE		3.3	5.3	7.3	2.7	4.8	6.7

Note: Tabulated values are $B^{R}(F)$, $A^{R}(F)$, and $R^{R}(F)$ as defined in Eqs. (7.12)–(7.14). Record length at each site, n, is 30 or 60.

which have record length $n_i = 30$ at each site, and for regions with the same frequency distributions as Regions R1–R4 but record length $n_i = 60$ at each site. In the homogeneous regions, R1 and R3, the pattern of the results is the same as for at-site analysis; doubling the record length at each site reduces the bias by a factor of approximately 2 and the RMSE by a factor of approximately $\sqrt{2}$. In the heterogeneous regions, R2 and R4, the patterns are less clear. The regional average absolute bias is mostly due to heterogeneity in the region and is not affected by increasing the record lengths. The RMSE of quantile estimators is less when $n = 60$, but by a factor of less than $\sqrt{2}$.

In heterogeneous regions, the presence of bias that does not vanish with increasing record lengths means that when record lengths are large, regional estimators can be less accurate than at-site estimators. This effect is illustrated in Figure 7.13, which shows the regional average relative RMSE of quantile estimators, both regional and at-site, in regions with the same frequency distributions as Region R2 but with different values of n, the record length at each site. For at-site estimators the relative RMSE decreases steadily, roughly proportionally to the $n_i^{-1/2}$ rate suggested by asymptotic theory. For regional estimators, bias due to heterogeneity imposes a lower bound below which the relative RMSE of quantile estimators cannot fall. When $n > 40$, at-site estimation is more accurate than regional estimation for the $F = 0.9$ quantile. Larger record lengths are needed for at-site estimation to be preferable at more extreme quantiles. This graph is, however, misleading in that

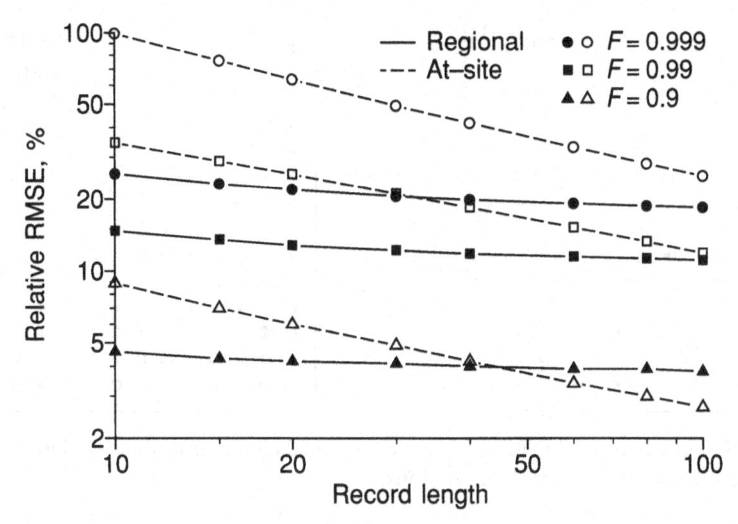

Fig. 7.13. Regional average relative RMSE of estimated quantiles for heterogeneous regions similar to Region R2 but with different record lengths. Fitted distribution: GEV.

when record lengths are large, heterogeneity in the region is more easily detected by the heterogeneity measures defined in Chapter 4. For realizations of Region R2 with $n = 60$, for example, the average value of the H statistic in Eq. (4.5) is 3.0, indicating that such a region is likely to be declared heterogeneous and to be subdivided into smaller regions, rather than being used itself in regional frequency analysis.

7.5.6 Intersite dependence

Regional frequency analysis assumes that data from different sites are statistically independent. In practice it is common for observations for the same time point at different sites to be positively correlated, with the correlation typically being higher for sites that are close to each other.

From a theoretical viewpoint, the effect of intersite dependence on the regional L-moment algorithm is to increase the variability of the regional averages (6.2) and (6.3). This increases the variability of the estimated growth curve, the last term of Eq. (7.3). It may also affect the bias of estimated quantiles through the last term of Eq. (7.4), though this effect is small.

Hosking and Wallis (1988) presented results of Monte Carlo simulation experiments designed to assess the effect of intersite dependence. We repeated some of their experiments for our representative regions. The procedure for generating correlated data is as given in Table 6.1 and involves specifying a matrix $\mathbf{R}$ whose typical element ρ_{ij} is the correlation between observations made at the same time point at sites i and j.

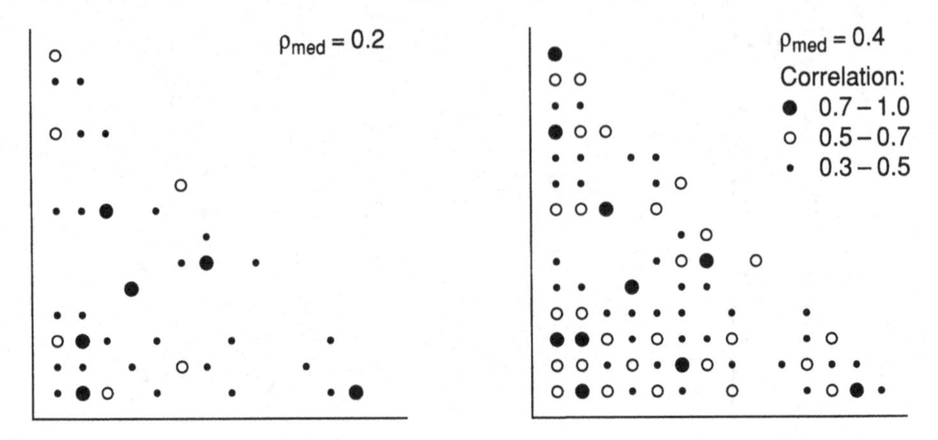

Fig. 7.14. Schematic representation of the off-diagonal elements of the correlation matrices used in the simulations of correlated regions.

In fact, as is clear from Table 6.1, ρ_{ij} is the correlation between data values that have been transformed to Normality, rather than of the data themselves. This is a matter of computational convenience and makes little practical difference for most moderately skew distributions. For example, for the sites in Regions R1–R4 the largest difference occurs for the site with highest skewness, site 15 in Region R2. If Normal variates with correlation 0.2 are transformed so that their distributions are generalized extreme-value with $\tau_3 = 0.3$ (corresponding to site 15 in Region R2), then the correlation of the transformed variates is 0.16; Normal correlations of 0.4, 0.6, and 0.8 transform to generalized extreme-value correlations of 0.34, 0.54, and 0.74, respectively.

To generate the population correlation coefficients, the N sites were chosen to be points randomly located, with uniform distribution, in the unit square. The correlation ρ_{ij} between sites i and j was set to $\exp(-\alpha d_{ij})$, where d_{ij} is the distance between sites i and j and α is chosen so that the median of the ρ_{ij}, $1 \le i < j \le N$, is equal to some specified value ρ_{med}. This method is not intended as a formal model of the variation of correlation with intersite distance, but merely as a means of achieving a realistic spread of ρ_{ij} values while ensuring that the correlation matrix is valid, that is, positive definite. The entire set of correlations is parametrized by the single value ρ_{med}, which determines a typical correlation coefficient and the total amount of dependence in the region.

Some typical correlation matrices generated by this procedure are shown in schematic form in Figure 7.14. These are the matrices used in our simulations of 15-site regions with ρ_{med} equal to 0.2 or 0.4. The pattern of correlation for $\rho_{\text{med}} = 0.2$ is quite similar to that of the sample correlations of British annual maximum streamflow data examined by Hosking and Wallis (1987b), suggesting that this procedure for generating correlation matrices does give realistic results.

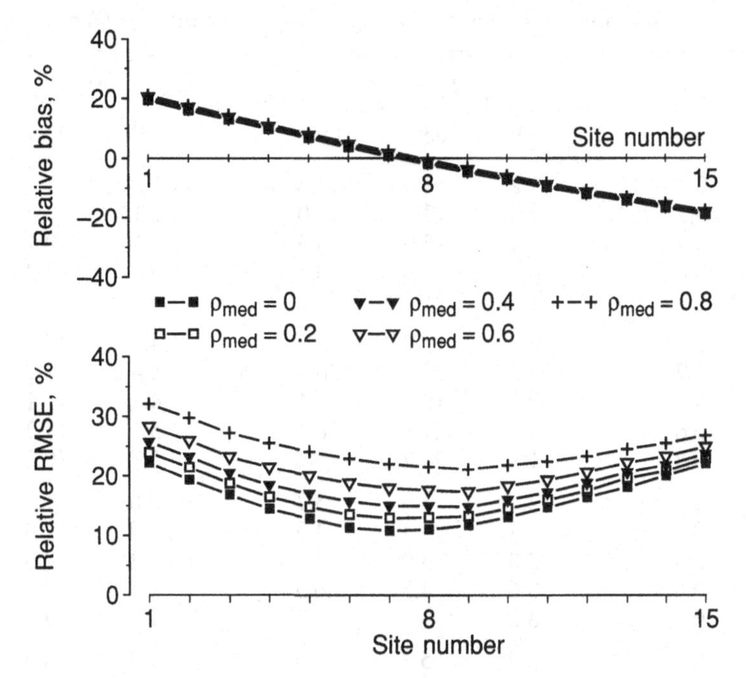

Fig. 7.15. Simulation results for correlated versions of Region R2. Graphs show the relative bias and relative RMSE of the estimator of the $F = 0.99$ quantile, $\hat{Q}_i(0.99)$.

The general effect of intersite dependence is illustrated by the simulation results for correlated versions of Region R2, shown in Figure 7.15. The bias is little affected by the presence of intersite dependence, but the RMSE of the estimated quantile increases steadily at all sites for the more correlated regions. The illustrated results are for the quantile estimator $\hat{Q}_i(0.99)$, but similar results were obtained for other quantile estimators and for the estimated growth curve.

The uniformity over different sites of the effect of intersite dependence means that the effect of intersite dependence can be well summarized, even for heterogeneous regions, by the regional average relative RMSE of the estimated quantiles and growth curve. Summary results for Regions R1–R4 are given in Table 7.9. The steady increase in RMSE values as the amount of correlation in the region increases, as measured by ρ_{med}, is apparent. For very highly dependent heterogeneous regions, at the less extreme quantiles, the RMSE of the regional estimators is higher than that of the at-site estimators, for which the corresponding results are the RMSE rows of Table 7.7. This implies that when data from different sites are highly correlated, heterogeneity is a major concern; regional frequency analysis should then be used only if there are strong reasons for believing that homogeneous regions can be identified. At lower levels of dependence, however, the effect of intersite

Table 7.9. *Simulation results for regions with intersite dependence.*

		Quantiles			Growth curve		
Region	ρ_{med} F:	0.9	0.99	0.999	0.9	0.99	0.999
	0	9.2	11.0	14.6	1.3	5.7	11.0
	0.2	9.6	13.0	19.0	2.1	8.0	15.3
R1	0.4	9.9	15.0	23.5	2.6	10.1	19.6
	0.6	10.4	17.8	30.3	3.2	12.8	25.9
	0.8	11.0	21.6	40.7	4.0	16.4	35.3
	0	10.2	15.7	23.0	4.1	12.2	20.6
	0.2	10.5	17.3	26.4	4.4	13.7	23.7
R2	0.4	10.9	18.9	30.1	4.7	15.1	27.1
	0.6	11.3	21.3	36.1	5.2	17.2	32.4
	0.8	11.8	24.7	45.6	5.7	20.2	40.9
	0	2.7	3.0	3.7	0.6	1.6	2.6
	0.2	2.7	3.4	4.5	0.9	2.3	3.6
R3	0.4	2.8	3.9	5.3	1.2	2.9	4.6
	0.6	3.0	4.5	6.5	1.5	3.7	6.0
	0.8	3.2	5.4	8.3	1.8	4.7	7.9
	0	3.3	5.3	7.3	1.9	4.5	6.7
	0.2	3.4	5.6	7.8	2.1	4.8	7.3
R4	0.4	3.5	5.9	8.4	2.2	5.2	7.9
	0.6	3.6	6.3	9.3	2.4	5.7	8.9
	0.8	3.7	7.1	10.7	2.6	6.5	10.3

Note: Tabulated values are the regional average relative RMSE of estimated quantiles, that is, $R^{R}(F)$ as defined in Eq. (7.14), and the corresponding quantities for the estimated growth curve.

dependence is fairly small, particularly for heterogeneous regions. For example, in regions similar to Region R2 where correlations between different sites do not typically exceed 0.2, the increase in the RMSE of quantile estimates should not exceed about 10% at the $F = 0.99$ quantile and 15% at the $F = 0.999$ quantile. This is comparable to the magnitude of the bias due to heterogeneity.

Because the main effect of intersite dependence is to increase the variability of estimators, it acts similarly to a reduction in the number of sites in the region. Its effect has sometimes been summarized by calculating an "effective number of independent sites," that is, the number of sites in a region comparable to the region of interest, but whose sites have mutually independent frequency distributions, such that the RMSEs of quantile estimators in the two regions are the same. Analytical formulas have been proposed for the effective number of sites (e.g., Alexander, 1954; Stedinger, 1983) though they seem to be seriously inaccurate in practice (Hosking and Wallis, 1988). Alternative methods of estimating an effective number

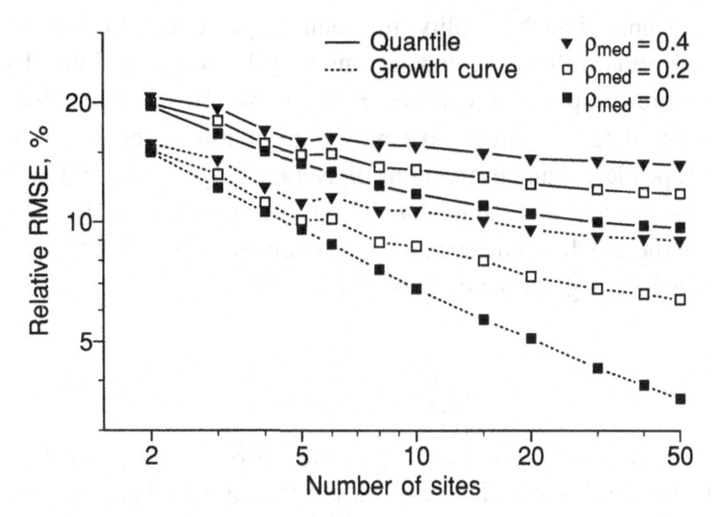

Fig. 7.16. Regional average relative RMSE of estimated quantiles and growth curve for homogeneous correlated regions similar to Region R1 but with different numbers of sites. Results are for nonexceedance probability 0.99. Record length 30 at each site. Fitted distribution: GEV.

of sites have been used by Dales and Reed (1989), Schaefer (1990), and Reed and Stewart (1994). Our results indicate that no single measure of an effective number of sites is uniformly applicable. The number depends on whether the region is homogeneous or heterogeneous, on whether estimation of quantiles or of the growth curve is being considered, and on which nonexceedance probability is of interest. For estimating the $F = 0.99$ quantile, for example, we find the effective number of independent sites in our representative regions when they are correlated with $\rho_{\text{med}} = 0.4$ to be about 4 for Region R1, between 5 and 6 for Region R2, between 4 and 5 for Region R3, and a little over 6 for Region R4.

A final consequence of intersite dependence is that, even in homogeneous regions, the accuracy with which the growth curve can be estimated does not decrease to zero as the number of sites in the region increases. After a certain point, the addition of sites with observations at the same time points as other sites in the region merely duplicates information already obtained from the other sites. The effect is illustrated in Figure 7.16 for estimators of the quantile $Q_i(0.99)$ and the growth factor $q_i(0.99)$ in regions analogous to Region R1 but with different numbers of sites. The results for $\rho_{\text{med}} = 0$ are the same as those for $F = 0.99$ in Figure 7.11. The results confirm that, in regional frequency analysis, little benefit should be expected from using more than about 20 sites in a region.

We emphasize that these conclusions arc valid for the particular dependence structure used in the simulations; that is, after transforming each site's frequency distribution to univariate Normality, the joint distributions at different sites are multivariate Normal. Different conclusions may be reached if the dependence structure is more complicated than this. For example, Hosking and Wallis (1988) simulated some data for which extremely high or low observations were more highly interdependent than observations of average magnitude and found that estimated quantiles were less accurate than for the "transformed multivariate Normal" dependence structure. It is, however, not easy to decide which dependence structure is appropriate for any given set of data.

7.5.7 Heterogeneity

The effect of heterogeneity is to increase the RMSE of the estimated quantiles and growth curve as a consequence of bias in the estimated growth curve at sites whose frequency distributions are different from the average for the entire region. The general pattern is exemplified by the results for Regions R2 and R4, shown in Tables 7.2 and 7.4 and Figures 7.3 and 7.5.

For a more general assessment of the effect of heterogeneity, we constructed regions analogous to Regions R2 and R4 but with different amounts of heterogeneity. Regions had 15 sites with record length 30 at each site. Frequency distributions at each site were specified by their L-CV τ and L-skewness τ_3, which varied linearly from site 1 to site 15. For regions analogous to Region R2, the frequency distribution at each site was generalized extreme-value, and site 8, the central site, had $\tau = 0.25$ and $\tau_3 = 0.25$, the same as the sites in Region R1. For regions analogous to Region R4, the frequency distribution at each site was lognormal, and site 8 had $\tau = 0.08$ and $\tau_3 = 0.05$, the same as the sites in Region R3. Quantiles and growth curves were estimated by the regional L-moment algorithm, the fitted distribution being correctly specified: generalized extreme-value for regions analogous to Region R2, lognormal for regions analogous to Region R4. Simulation results are summarized in Table 7.10 by the regional average relative RMSE of quantile and growth curve estimators. The table also shows the average value, over 1,000 simulated realizations of each region, of the heterogeneity measure H defined in Eq. (4.5). As the amount of heterogeneity, measured by the range of τ or τ_3 across the region, increases, the average value of H and the RMSE of estimated quantiles and growth curve increase uniformly.

The relative RMSE values in Table 7.10 can be compared with the corresponding values for at-site estimation. The ratio

$$\frac{R^{\mathrm{R}}(F) \text{ for at-site estimation}}{R^{\mathrm{R}}(F) \text{ for regional estimation}} \tag{7.15}$$

Table 7.10. *Simulation results for heterogeneous regions.*

Average		Range			Ave.		Quantile			Growth curve		
τ	τ_3	τ	τ_3	Dist.	H	F:	0.9	0.99	0.999	0.9	0.99	0.999
.25	.25	.00	.00	GEV	0.00		9.2	11.0	14.6	1.3	5.7	11.0
.25	.25	.05	.05	GEV	0.60		9.5	12.4	17.2	2.4	7.9	14.2
.25	.25	.10	.10	GEV	1.78		10.2	15.7	23.0	4.1	12.2	20.6
.25	.25	.15	.15	GEV	3.41		11.3	19.8	30.1	5.9	17.0	28.2
.25	.25	.20	.20	GEV	5.24		12.7	24.4	37.9	7.7	22.0	36.3
.25	.25	.25	.25	GEV	7.17		14.3	29.4	46.2	9.6	27.3	44.9
.08	.05	.00	.000	LN3	0.00		2.7	3.0	3.7	0.6	1.6	2.6
.08	.05	.01	.033	LN3	0.22		2.7	3.4	4.3	0.9	2.2	3.4
.08	.05	.02	.067	LN3	0.84		3.0	4.3	5.7	1.4	3.3	5.0
.08	.05	.03	.100	LN3	1.78		3.3	5.3	7.3	1.9	4.5	6.7
.08	.05	.04	.133	LN3	2.93		3.7	6.5	9.1	2.5	5.8	8.6
.08	.05	.05	.167	LN3	4.21		4.2	7.7	10.9	3.0	7.1	10.5

Note: All regions have 15 sites and record length 30 at each site. Frequency distributions have L-CV and L-skewness varying linearly from site 1 to site 15; average and range of these values are tabulated. "Dist." is the frequency distribution at each site: generalized extreme-value (GEV) or lognormal (LN3). "Ave. H" is the average value, over simulated realizations of the region, of the H statistic defined in Eq. (4.5). Tabulated values are the regional average relative RMSE of estimated quantiles, that is, $R^{\mathrm{R}}(F)$ as defined in Eq. (7.14), and the corresponding quantities for the estimated growth curve.

measures the relative accuracy of the two estimation methods. Values larger than 1 indicate that regional estimation is more accurate. This ratio is plotted as a function of the heterogeneity measure H for regions analogous to Region R2 in Figure 7.17 and for regions analogous to Region R4 in Figure 7.18. As the amount of heterogeneity, as measured by H, increases, the advantage of regional estimation over at-site estimation decreases. In very heterogeneous regions, as H increases beyond a certain point, at-site estimation becomes preferable. This point is reached more rapidly for quantiles in the body of the distribution; for extreme quantiles, regional estimation remains preferable even when H is large. For regions with low skewness, analogous to Region R4, at-site estimation becomes preferable at lower values of H.

For the regions in Figures 7.17 and 7.18, regional estimation is more accurate than at-site estimation for all quantiles with $F \geq 0.9$ when the average H value is less than 2. For the most extreme quantiles, regional estimation remains preferable even in regions for which the average H value is considerably larger than 2. This may suggest that the criterion $H < 2$ proposed in Chapter 4 for declaring a region to be "definitely heterogeneous" is too strict. However, there are other possibilities than regional analysis of all sites in a region; it may be possible to subdivide the region

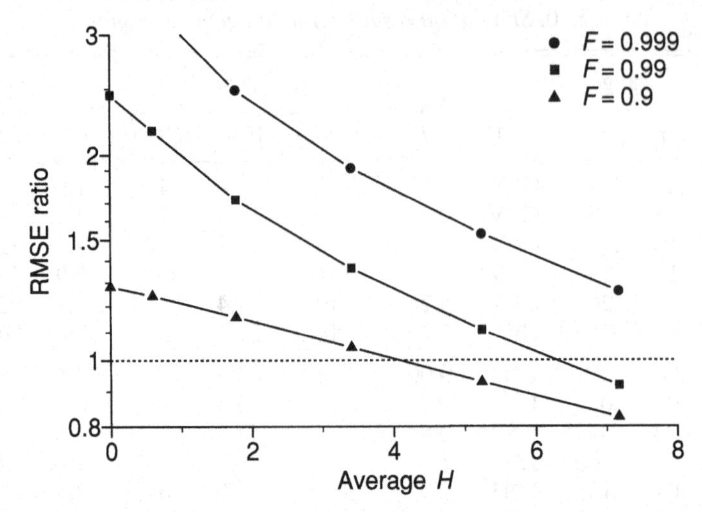

Fig. 7.17. Ratio of the RMSE of at-site quantile estimators to regional quantile estimators for heterogeneous regions "analogous to Region R2" (as defined in text). Record length 30 at each site. Fitted distribution: GEV.

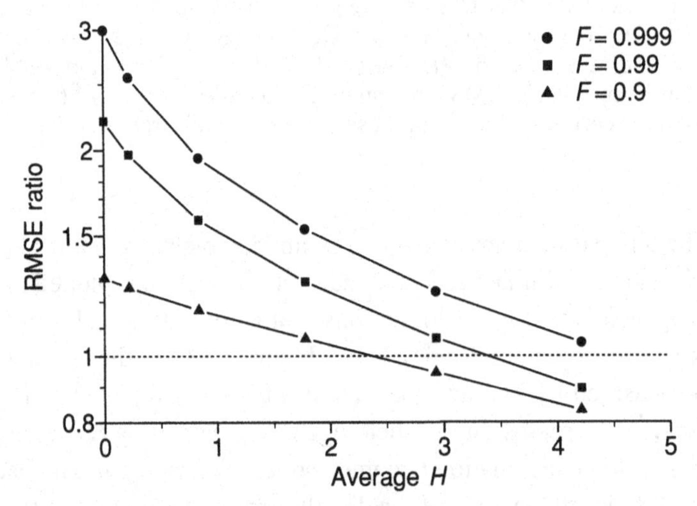

Fig. 7.18. Ratio of the RMSE of at-site quantile estimators to regional quantile estimators for heterogeneous regions "analogous to Region R4" (as defined in text). Record length 30 at each site. Fitted distribution: lognormal.

into smaller regions or by other means revise the assignment of sites to regions. We feel that when an H value greater than 2 is observed, these other possibilities merit investigation.

We have so far considered heterogeneous regions in which L-CV and L-skewness both vary from site to site. In principle, heterogeneity may arise from differences

Table 7.11. *Simulation results for regions with heterogeneity in τ and τ_3.*

Range			Quantile ratio			Relative RMSE			Average H		
τ	τ_3	F:	0.9	0.99	0.999	0.9	0.99	0.999	V	V_2	V_3
.10	.10		1.15	**1.48**	1.88	10.2	**15.7**	23.0	1.74	0.54	0.24
.15	.00		1.25	**1.47**	1.59	11.4	**15.6**	19.7	3.43	0.88	0.10
.00	.30		0.97	**1.48**	2.60	9.7	**16.1**	30.1	0.51	1.56	1.32
.10	.10		1.15	1.48	**1.88**	10.2	15.7	**23.0**	1.74	0.54	0.24
.20	.00		1.35	1.68	**1.88**	12.8	18.5	**23.2**	5.30	1.39	0.10
.00	.20		0.98	1.31	**1.91**	9.4	13.6	**23.3**	0.32	0.78	0.66

Note: All regions have 15 sites with generalized extreme-value frequency distributions and record length 30 at each site. Frequency distributions have L-CV τ and L-skewness τ_3 varying linearly from site 1 to site 15, the range of these values is tabulated, and average values are $\tau = 0.25$ and $\tau_3 = 0.25$. "Quantile ratio" is the ratio of the quantile at site 15 to that at site 1, that is, $Q_{15}(F)/Q_1(F)$. "Relative RMSE" is the regional average relative RMSE of estimated quantiles, that is, $R^R(F)$ as defined in Eq. (7.14). "Average H" is the average value, over simulated realizations of the region, of the three H statistics defined in Eq. (4.5) and based on the heterogeneity measures (4.4), (4.6), and (4.7).

in L-CV, in L-skewness, or in any other feature or combination of features of the frequency distribution. Though any kind of heterogeneity increases the bias of the estimated growth curve in regional frequency analysis, different kinds of heterogeneity have different effects for estimation of any given quantile and may not be easy to detect in practice.

As an example, we compare the effects of heterogeneity in L-CV alone, L-skewness alone, and L-CV and L-skewness together. Region R2 is an example of the last kind, with L-CV and L-skewness both having a range of 0.10 between the most extreme sites of the region. For this region the growth curve at nonexceedance probability 0.99, $q(0.99)$, is 48% larger at site 15 than at site 1. To achieve this ratio of $q(0.99)$ values in a region in which only L-CV varies, the range of L-CV would need to be 0.15; and to achieve the same ratio in a region in which only L-skewness varies, the range of L-skewness would need to be 0.30. In each case, the regions are "analogous to Region R2": 15 sites, generalized extreme-value frequency distributions at each site, average L-CV and L-skewness both 0.25, and L-CV or L-skewness varying linearly from site 1 to site 15. Regions were similarly constructed to achieve, through variations in L-CV or L-skewness alone, equal quantile ratios for the $F = 0.999$ quantile. Simulation results for these regions are summarized in Table 7.11.

In Table 7.11 the bold numbers show that equal quantile ratios for a given quantile correspond well to equal RMSEs for the estimator of the quantile. However, the ease with which heterogeneity can be detected, as measured by the "Average H" values in Table 7.11, is not consistent across regions that have the same RMSE of their estimated quantiles. In particular, heterogeneity in L-skewness alone, which can have a serious effect on the accuracy of estimated quantiles in the extreme tail of the distribution, is not easily detected even by the heterogeneity measures based on V_2 and V_3, which explicitly involve between-site variation in sample L-skewness statistics. In practice this should not be a serious problem, for it is hard to imagine circumstances in which L-skewness varies from site to site but L-CV does not. When heterogeneity in L-CV is sufficient to affect the accuracy of quantile estimators, there should be a good chance that it can be detected in regions where the typical record length is 30 or more. However, the general conclusion must be that no single heterogeneity measure is likely to give a complete picture of the amount of heterogeneity in a region.

7.5.8 Misspecification of the regional frequency distribution

Misspecification of the frequency distribution limits the accuracy of estimators of the quantiles and growth curve. A good fit to part of the distribution can be achieved, but bias is increasingly apparent in estimated quantiles in the tails of the distribution. This bias affects estimated quantiles at all sites, and is reflected in the regional average bias measure $B^{\mathrm{R}}(F)$ defined in Eq. (7.12); unlike bias due to heterogeneity, it does not cancel out when averaged across the sites in the region.

Several distributions, some misspecified, were fitted to data simulated from Region R1. Regional average performance measures for quantile estimators are given in Table 7.12.

Results for regional fitting of three-parameter distributions show a consistent pattern. The three-parameter distributions in Table 7.12 – the middle block of rows – are listed in decreasing order of their L-kurtosis when their L-skewness is 0.25, the true value for Region R1. Because the fitted distributions all have the same L-CV and L-skewness, the distributions with higher L-kurtosis have higher quantiles in the extreme upper tail of the distribution. This is reflected in the bias values for estimators of the $F = 0.999$ quantile; the biases are ordered in the same way as the distributions' L-kurtosis. For severely misspecified distributions, bias is the main contributor to the regional average RMSE. This is particularly true of the generalized Pareto distribution, which has L-kurtosis very different from that of the generalized extreme-value distribution when $\tau_3 = 0.25$ (see Figure A.2 in the appendix). For the other three-parameter distributions, misspecification is important only in the extreme tail of the distribution, and it has little effect on the RMSE of estimated quantiles for $F \leq 0.99$.

Table 7.12. *Simulation results for different fitted distributions: Region R1.*

Fitted distribution	Pars.	$F = 0.9$		$F = 0.99$		$F = 0.999$	
		Bias	RMSE	Bias	RMSE	Bias	RMSE
Gumbel(AS)	2	0.4	12.7	−10.4	17.2	−22.1	25.6
GEV(AS)	3	−0.8	11.8	0.6	26.9	7.8	56.8
Gumbel	2	0.2	9.3	−10.6	13.6	−22.4	23.6
G. logistic	3	−2.5	9.3	2.1	11.2	16.0	22.7
GEV	3	−0.2	9.2	−2.0	11.0	−3.9	14.6
Lognormal	3	0.6	9.4	−3.7	11.0	−9.6	15.1
Pearson III	3	1.9	9.7	−6.9	11.9	−18.9	21.0
G. Pareto	3	4.1	10.5	−12.8	16.2	−33.0	34.4
Kappa	4	−0.2	9.3	−2.1	11.3	−3.5	17.2
Wakeby	5	0.4	9.4	−2.5	11.5	−8.4	18.6

Note: All regions have 15 sites with generalized extreme-value frequency distributions and record length 30 at each site. Frequency distributions have *L*-CV $\tau = 0.25$ and *L*-skewness $\tau_3 = 0.25$. AS denotes at-site estimation; otherwise estimation is by the regional *L*-moment algorithm. "Pars." is the number of distribution parameters that must be estimated. "Bias" and "RMSE" are the regional average relative bias and RMSE of estimated quantiles, that is, $B^{\mathrm{R}}(F)$ and $R^{\mathrm{R}}(F)$ as defined in Eqs. (7.12) and (7.14).

Table 7.12 also illustrates comparisons between fitted distributions with different numbers of parameters. The Gumbel distribution has two parameters, and estimated quantiles derived from it tend to have less variability than those of three-parameter distributions. However, the distribution has fixed *L*-skewness of 0.1699, less than the *L*-skewness of the sites in Region R1. In consequence, the fitted Gumbel distribution has large negative bias at extreme quantiles. In practice, the use of a two-parameter distribution in regional frequency analysis is likely to be beneficial only if the investigator has complete confidence that the distribution's *L*-skewness and *L*-kurtosis are close to those of the frequency distributions at the sites in the region.

Distributions with more parameters tend to yield estimated quantiles that are more variable. This can be seen by comparing the results in Table 7.12 for the generalized extreme-value and kappa distributions, which have three and four parameters, respectively. Both distributions are correctly specified for Region R1 and therefore have low bias, but the RMSEs of estimated quantiles are somewhat higher for the kappa distribution. The Wakeby distribution is misspecified, but it has many parameters enabling it to mimic the shape of many other distributions over a fairly wide range of quantile values. Its performance measures for quantile estimation are only a little worse than those of the kappa distribution.

Table 7.13. *Simulation results for different fitted distributions: Region R2.*

		$F = 0.9$			$F = 0.99$			$F = 0.999$		
Fitted distribution	Pars.	Bias	Abs. bias	RMSE	Bias	Abs. bias	RMSE	Bias	Abs. bias	RMSE
Gumbel(AS)	2	0.5	0.5	12.8	−10.4	10.4	17.4	−21.9	21.9	25.8
GEV(AS)	3	−0.8	0.8	11.8	0.6	0.6	27.0	7.7	7.7	57.2
Gumbel	2	0.4	3.7	10.3	−9.9	12.2	15.7	−20.9	22.0	24.1
G. logistic	3	−2.2	4.1	10.1	2.9	10.8	16.2	18.1	23.1	29.9
GEV	3	0.1	3.7	10.2	−1.2	10.4	15.7	−2.1	16.6	23.0
Lognormal	3	0.9	3.8	10.4	−2.9	10.5	15.4	−8.0	16.9	21.7
Pearson III	3	2.1	4.0	10.8	−6.1	11.0	15.4	−17.4	19.9	23.1
G. Pareto	3	4.4	5.0	11.6	−12.1	13.4	17.7	−31.8	31.8	33.8
Kappa	4	0.1	3.7	10.3	−1.3	10.4	15.9	−1.5	16.6	25.1
Wakeby	5	0.6	3.8	10.4	−1.6	10.3	16.0	−6.5	16.7	25.2

Note: Region specification as in Table 7.1. AS denotes at-site estimation; otherwise estimation is by the regional *L*-moment algorithm. "Bias," "Abs. bias," and "RMSE" are the regional average relative bias, absolute bias and RMSE of estimated quantiles, that is, $B^{R}(F)$, $A^{R}(F)$, and $R^{R}(F)$ as defined in Eqs. (7.12)–(7.14).

Results for at-site estimation are also included in Table 7.12, but merely confirm that at-site estimation is not competitive for homogeneous regions.

Analogous simulation results for the heterogeneous Region R2 are given in Table 7.13. As before, the effect of heterogeneity is to add a bias to the estimated quantiles that varies from site to site but cancels out when averaged over the entire region. This is reflected in the "Bias" columns in Table 7.13, which are little changed from those of Table 7.12. Heterogeneity increases the regional average absolute bias and RMSE values.

An important interaction between heterogeneity and misspecification is that in heterogeneous regions, misspecification has less effect on the RMSE of estimated quantiles. In a heterogeneous region, no distribution can be correctly specified at every site, but for Region R2, in which all sites have generalized extreme-value frequency distributions, the generalized extreme-value distribution and its generalization, the kappa distribution, come closest to being correctly specified and have the lowest bias values. However, the generalized extreme-value distribution does not have uniformly the lowest RMSE of estimated quantiles; the lognormal distribution has lower RMSE for $F = 0.99$ and $F = 0.999$. This suggests that the lognormal distribution has robustness properties, in that its performance is competitive even when it is a clear misspecification of the true distribution. Comparison of the results in Tables 7.12 and 7.13 for the generalized extreme-value, Gumbel, and Pearson type III distributions also shows that the Gumbel and Pearson type III distributions are more nearly competitive with the generalized extreme-value distribution for the heterogeneous region, though they retain considerable negative bias in their estimated quantiles.

7.5.9 Heterogeneity and misspecification

As a wider study of misspecification and robustness, variants of Region R2 were constructed that had different true frequency distributions. The regions each had 15 sites with record length 30 at each site; *L*-CV and *L*-skewness varied linearly from 0.2 at site 1 to 0.3 at site 15, and each site had a frequency distribution of the same kind, which could be generalized logistic, generalized extreme-value, lognormal or Pearson type III. The regional *L*-moment algorithm was used to fit each of these four three-parameter distributions, together with the kappa and Wakeby distributions. At-site estimation using the correctly specified distribution was also carried out. The regional average performance measures are given in Table 7.14. The general pattern of the results is similar for each true distribution. When the fitted distribution is misspecified, having higher or lower *L*-kurtosis than the true distribution, positive or negative bias and an increase in RMSE are observed in the extreme upper quantiles. For these regions the effects are not serious except in cases of severe misspecification or when quantiles with nonexceedance probability $F > 0.99$ are to be estimated; in regions with less heterogeneity or longer at-site record lengths, bias due to misspecification would be more important. The kappa and Wakeby distributions generally give estimated quantiles with no severe bias and RMSEs only a little larger than when the fitted distribution is correctly specified. They are therefore robust to misspecification and should be a good choice in regions for which the goodness-of-fit measure described in Chapter 5 does not yield a clear choice of an appropriate distribution to fit to the data. The lognormal distribution is more robust than the other three-parameter distributions considered here, though it gives large biases when fitted to data drawn from a generalized Pareto distribution. Similar results were obtained for regions in which the frequency distributions have lower skewness, analogous to Region R4.

From our simulations we judge that the kappa and Wakeby distributions, and to a lesser extent the lognormal distribution, have good robustness with respect to misspecification of the frequency distribution. These general conclusions should be valid for a wide range of regions that have frequency distributions similar to those in Figures 7.1 or 7.4. Different conclusions may of course apply in other situations, such as when frequency distributions have bimodal probability density functions or a significant proportion of exact zero values.

Table 7.14. *Simulation results for different combinations of true and fitted distributions.*

Fitted distribution	Bias, True distribution					RMSE, True distribution				
	GLO	GEV	LN3	PE3	GPA	GLO	GEV	LN3	PE3	GPA
					$F = 0.9$					
At-site	1.1	−0.8	−0.5	−2.9	−6.3	12.7	11.8	12.1	12.1	11.3
G. logistic	0.0	−2.2	−3.1	−4.5	−6.3	10.7	10.1	10.2	10.5	10.8
GEV	2.4	0.1	−0.8	−2.2	−4.1	11.3	10.2	10.1	10.1	10.1
Lognormal	3.1	0.9	0.0	−1.4	−3.3	11.6	10.4	10.2	10.2	9.9
Pearson III	4.3	2.1	1.3	−0.1	−1.9	12.2	10.8	10.5	10.3	9.8
G. Pareto	6.8	4.4	3.5	2.1	0.1	13.3	11.6	11.2	10.7	9.9
Kappa	1.3	0.1	0.0	0.0	0.0	11.0	10.3	10.3	10.3	10.0
Wakeby	1.2	0.6	0.4	0.0	−0.2	10.9	10.4	10.4	10.4	9.9
					$F = 0.99$					
At-site	−3.2	0.6	0.9	6.2	6.2	32.3	27.0	24.1	24.1	23.4
G. logistic	−1.9	2.9	5.2	9.5	16.8	16.0	16.2	16.3	17.2	22.2
GEV	−6.0	−1.2	1.0	5.2	12.3	16.3	15.7	15.3	15.4	19.4
Lognormal	−7.5	−2.9	−0.8	3.3	10.2	16.4	15.4	14.9	14.7	18.1
Pearson III	−10.4	−6.1	−4.1	−0.2	6.3	16.9	15.4	14.6	13.8	16.3
G. Pareto	−16.5	−12.1	−10.1	−6.3	0.1	20.6	17.7	16.4	15.0	15.0
Kappa	−3.2	−1.3	−0.8	−0.3	0.6	16.3	15.9	15.3	14.3	15.0
Wakeby	−2.7	−1.6	−1.1	−0.4	0.8	16.7	16.0	15.3	14.3	15.1
					$F = 0.999$					
At-site	−7.4	7.7	4.2	4.2	27.8	68.1	57.2	39.7	39.7	57.2
G. logistic	−3.8	18.1	27.1	43.8	73.6	23.1	29.9	34.1	47.6	76.4
GEV	−20.5	−2.1	5.4	19.4	44.4	27.0	23.0	22.1	26.9	48.4
Lognormal	−24.8	−8.0	−1.1	11.7	34.7	28.4	21.7	19.6	21.0	39.1
Pearson III	−32.1	−17.4	−11.4	−0.1	20.0	33.6	23.1	19.4	16.2	27.4
G. Pareto	−44.5	−31.8	−26.6	−16.9	0.4	45.5	33.8	29.1	20.2	20.2
Kappa	−9.4	−1.5	−0.9	−0.6	1.9	25.3	25.1	22.4	19.2	21.0
Wakeby	−13.4	−6.5	−4.5	−0.4	4.3	27.7	25.2	22.7	20.3	22.6

Note: All regions have 15 sites with record length 30 at each site. Frequency distributions have *L*-CV τ and *L*-skewness τ_3 varying linearly from 0.2 at site 1 to 0.3 at site 15 and are of type generalized logistic (GLO), generalized extreme-value (GEV), lognormal (LN3), Pearson type III (PE3) or generalized Pareto (GPA). "Bias" and "RMSE" are the regional average relative bias and RMSE of estimated quantiles, that is, $B^R(F)$ and $R^R(F)$ as defined in Eqs. (7.12) and (7.14).

7.5.10 Heterogeneity, misspecification, and intersite dependence

Our simulation experiments have shown that intersite dependence, heterogeneity, and misspecification of the frequency distribution can significantly degrade the accuracy of the estimated quantiles and growth curve obtained through the regional

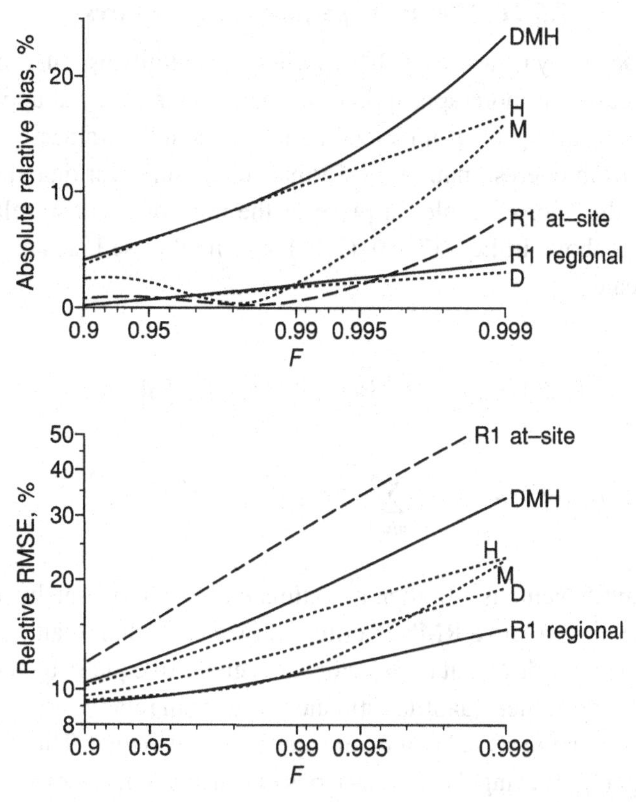

Fig. 7.19. Regional average relative RMSE of estimated quantiles for Region R1 (at-site and regional estimation) and variants of Region R1 containing intersite dependence (D), misspecification (M), heterogeneity (H), and all three (DMH).

L-moment algorithm. The effects of these factors, separately and together, are illustrated in Figure 7.19.

The base region is Region R1, homogeneous with a GEV frequency distribution and no intersite dependence. Figure 7.19 shows simulation results for quantile estimation for this region using both at-site and regional estimators. Figure 7.19 also shows results for regions with moderate amounts of intersite dependence ($\rho_{\text{med}} = 0.2$, in the notation of Subsection 7.5.6), misspecification (fitted distribution generalized logistic), and heterogeneity (region specification as for Region R2, given in Table 7.1), separately and together. Though each violation of the assumptions of the index-flood procedure reduces the accuracy of quantile estimation, even the combination of all three violations still yields estimated quantiles that are considerably more accurate than at-site quantile estimates that use the correctly specified frequency distribution.

7.5.11 Alternative performance measures

Although the accuracy measures (7.10)–(7.14) are generally useful, other measures may in some cases be more appropriate. In many environmental applications the economic loss resulting from underestimation of quantile values is greater than that resulting from overestimation. Performance measures for quantile estimators should reflect this. For example, in place of the relative bias and relative RMSE criteria defined above in Eqs. (7.10)–(7.11), one might use bias and RMSE on a logarithmic scale:

$$BL_i(F) = M^{-1} \sum_{m=1}^{M} \log\{\hat{Q}_i^{[m]}(F)/Q_i(F)\}, \tag{7.16}$$

$$RL_i(F) = \left(M^{-1} \sum_{m=1}^{M} [\log\{\hat{Q}_i^{[m]}(F)/Q_i(F)\}]^2\right)^{1/2}. \tag{7.17}$$

If the true quantile value is 1.0, then an estimate of 2.0 (too large by a factor of 2) is penalized by the "relative RMSE" criterion twice as much as an estimate of 0.5 (too small by a factor of 2), but these errors are penalized equally by the "RMSE on log scale" criterion. When quantile estimates have small relative errors the "relative error" and "error on log scale" criteria are approximately equal. This can be seen by expanding $\log(\hat{Q}/Q)$ using the Taylor-series expansion $\log(1+x) \approx x$ for small x, accurate up to a term of order x^2. We have

$$\log(\hat{Q}/Q) = \log\left(1 + \frac{\hat{Q} - Q}{Q}\right) \approx \frac{\hat{Q} - Q}{Q}, \tag{7.18}$$

whence it follows that $BL_i(F) \approx B_i(F)$ and $RL_i(F) \approx R_i(F)$.

The simulation results presented in the foregoing sections are generally not sensitive to the use of $B_i(F)$ and $R_i(F)$ instead of $BL_i(F)$ and $RL_i(F)$ as performance measures, but there are exceptions. When assessing the average performance across a region of different estimation procedures, differences are apparent when the regional average bias is substantial. Of the factors explored in our simulations, only misspecification of the frequency distribution has a major effect on the regional average bias. As an example, Table 7.15 contains a summary of the results of fitting different distributions to data simulated from Region R2. These results should be compared with those in Table 7.13 on page 136, which are for the same simulated data and the same estimators, but use the "relative error" criteria; Table 7.15 gives the results using the "error on log scale" criteria. To facilitate the comparison, the values tabulated in Table 7.15 are expressed as percentages.

Table 7.15. *Simulation results for Region R2, using log-scale performance measures.*

		$F = 0.9$			$F = 0.99$			$F = 0.999$		
Fitted distribution	Pars.	Bias	Abs. bias	RMSE	Bias	Abs. bias	RMSE	Bias	Abs. bias	RMSE
Gumbel(AS)	2	–0.3	0.3	12.5	–12.2	12.2	19.6	–26.7	26.7	31.7
GEV(AS)	3	–1.6	1.6	11.9	–2.8	2.8	25.4	–3.2	3.2	44.3
Gumbel	2	–0.1	3.8	10.3	–11.6	13.8	17.5	–25.9	26.8	29.3
G. logistic	3	–2.8	4.4	10.4	1.6	10.6	15.8	13.7	19.7	25.8
GEV	3	–0.4	3.9	10.2	–2.5	10.7	16.0	–5.1	17.4	23.8
Lognormal	3	0.3	3.8	10.3	–4.3	11.0	16.1	–11.1	18.9	23.8
Pearson III	3	1.6	4.0	10.5	–7.6	12.0	16.6	–21.7	23.9	27.5
G. Pareto	3	3.8	4.7	11.0	–14.3	15.4	20.1	–41.2	41.2	44.1
Kappa	4	–0.4	3.9	10.3	–2.6	10.7	16.3	–5.0	17.4	25.7
Wakeby	5	0.1	3.8	10.3	–3.0	10.8	16.4	–10.2	18.6	27.0

Note: Region specification as in Table 7.1. AS denotes at-site estimation; otherwise estimation is by the regional L-moment algorithm. Tabulated values are the log-scale performance measures $BL^{R}(F)$, $AL^{R}(F)$, and $RL^{R}(F)$, defined analogously to Eqs. (7.12)–(7.14) as averages of $BL_i(F)$, $|BL_i(F)|$, and $RL_i(F)$, respectively.

Distributions that have higher L-kurtosis than the true frequency distribution perform better according to the "error on log scale" criteria than according to the "relative error" criteria: they tend to yield estimated quantiles that are too high and which are penalized less by the "error on log scale" criteria. Thus the generalized logistic distribution performs approximately as well as does the true generalized extreme-value distribution for fitting to Region R2 according to the "error on log scale" criteria. Distributions that have lower L-kurtosis than the true frequency distribution perform correspondingly worse according to the "error on log scale" criteria: the Gumbel and Pearson type III distributions are no longer competitive with the best-fitting distributions for Region R2, and the generalized Pareto distribution performs poorly, just as in Table 7.13.

7.6 Summary of simulation results

Our simulation experiments have explored the performance of the regional L-moment algorithm under a wide range of conditions. The following general conclusions seem to be valid for a wide variety of homogeneous and moderately heterogeneous regions with a range of at-site frequency distributions.

- Regionalization is valuable. Even in regions with moderate amounts of heterogeneity, intersite dependence, and misspecification of the frequency distribution, regional frequency analysis is more accurate than at-site frequency analysis.
- Regionalization is particularly valuable for estimation of the growth curve and quantiles far into the tail of the frequency distribution. Error in estimation of quantiles in the main body of the frequency distribution is dominated by variability of the estimated index flood, which is not explicitly addressed by regionalization using an index-flood procedure.
- Errors in quantile estimates, and errors in growth curve estimates for heterogeneous regions, decrease fairly slowly as a function of the number of sites in the region when the number of sites is large. In consequence, there is often little gain in accuracy from using regions containing more than about 20 sites.
- Longer records make regional estimation less valuable relative to at-site estimation. However, heterogeneity is easier to detect when records are longer. This suggests that regions should contain fewer sites when at-site record lengths are large.
- We do not recommend the use of two-parameter distributions such as the Gumbel in regional frequency analysis. Use of a two-parameter distribution is beneficial only if the investigator has complete confidence that the distribution's *L*-skewness and *L*-kurtosis are close to those of the frequency distributions at the sites in the region; otherwise large biases in quantile estimates will result.
- Misspecification of the frequency distribution is important only for quantiles far into the tails of the distribution. In the upper tail, for example, for the representative regions of Section 7.4, misspecification is important only when $F > 0.99$.
- Certain robust distributions, such as the kappa and Wakeby, yield reasonably accurate estimates over a wide range of at-site frequency distributions for the region.
- Heterogeneity introduces bias into estimates at sites not typical of the region as a whole. This bias can easily be the major source of error in the estimated quantiles and growth curve.
- Intersite dependence increases the variability of estimates but has little effect on their bias. Small amounts of intersite dependence should not be a concern in regional estimation.
- At extreme quantiles ($F \geq 0.999$) the advantage of regional over at-site estimation is greater. For these quantiles, heterogeneity is less important as a source of error, whereas misspecification of the frequency distribution is more important.

These conclusions are valid for a range of regions similar to the representative regions defined in Section 7.4. These regions have frequency distributions that are unimodal with *L*-CV and *L*-skewness in the range 0 to 0.3. These regions should be representative of a variety of environmental applications, but we do not claim that they exhaustively cover the range of applications for which regional frequency analysis might be considered. Furthermore, some details of the performance of the regional *L*-moment algorithm depend on particular choices of the criteria used to measure the performance of estimators and on the particular specification of intersite dependence.

8
Other topics

8.1 Variants of regionalization

In this monograph we have concentrated on regional frequency analysis using the index-flood procedure defined in Section 1.3 and the comparison of this method with at-site estimation. Several other regional frequency analysis procedures have been proposed; here we briefly describe them. For simplicity we consider them in the context of estimating a frequency distribution parametrized by its mean, its dispersion divided by its mean (typically L-CV), and one or more shape parameters (typically L-skewness). The estimators that each regional frequency analysis procedure uses for the parameters are summarized in Table 8.1. We are concerned with the question of which data are used in the analysis: at-site, regional, or some combination of the two. We do not consider the question of which statistical methods to apply to the data. We believe that methods based on L-moments are the best currently available; other approaches are reviewed by Cunnane (1988).

At-site estimation

For reference, we note here that at-site estimation involves the use of at-site estimates for all of the parameters of the distribution.

Regional shape estimation

If the mean and dispersion are estimated from at-site statistics, and the shape parameters are estimated by averaging the at-site shape measures for the sites in a region, we call the resulting procedure a "regional shape estimation" procedure. It is intermediate between pure at-site estimation and the index-flood procedure. It is discussed in more detail in Section 8.2.

Table 8.1. *Estimates of distribution parameters used by different variants of regional frequency analysis.*

Variant	Mean	Dispersion	Shape
At-site	at-site	at-site	at-site
Regional shape estimation	at-site	at-site	regional average
Index-flood	at-site	regional average	regional average
Hierarchical regions	at-site	regional average for subregion	regional average for full region
Fractional membership	at-site	weighted average of regional estimates	
Region of influence	at-site	weighted average of at-site estimates, for sites in site i's region of influence	
Mapping	at-site	estimated function of site characteristics	
Bulletin 17	at-site	at-site	weighted average of at-site estimate and estimated function of site location

Index-flood procedure

When the dispersion and shape parameters are both estimated by regional averaging, while the mean is still estimated from at-site data, the index-flood procedure is obtained.

Hierarchical regions

The index-flood procedure uses the same region as the basis for estimating both dispersion and shape parameters. Fiorentino et al. (1987) and Gabriele and Arnell (1991) proposed a procedure that involved a hierarchy of regions. Relatively large regions are defined over which the shape parameters are assumed to be constant, and these regions are subdivided into smaller regions over which the dispersion parameter is assumed to be constant.

The foregoing procedures have the disadvantage that estimated parameters and quantiles may change abruptly when passing from one site to a neighboring site that has been assigned to a different region. Several regionalization procedures aim for a smooth transition between sites.

Fractional membership

A site may be regarded not as belonging to a particular region but as having fractional membership in several regions. Parameters or quantiles for the site can then be estimated by a weighted average of the corresponding estimates for different regions. Weights can be obtained by using discriminant analysis to relate the different regions to the site characteristics. If the weights are functions of site characteristics and do not involve the at-site data, estimates can be obtained even at sites where no data are available. Wiltshire (1986c) proposed this approach for estimation of quantiles at ungaged sites. It can equally well be used for estimation of distribution parameters such as τ or τ_3 and for estimation at gaged sites. Acreman and Sinclair (1986) used a clustering method that explicitly yields weights for the fractional membership of a gaged site in more than one region.

The use of fractional membership does not enable any relaxation of the criteria for identifying homogeneous regions. Even though there may be doubt about which region a site belongs to, the regions themselves should still be homogeneous and distinct. This is to ensure that the bias and variability of each region's parameter or quantile estimates, and thence of the final estimator averaged across different regions, are as small as possible.

When a smooth transition between regions is required, the use of regionalization with a fractional membership procedure is attractive. There may, however, be problems with the derivation of the weights for fractional membership. Discriminant analysis is usually based on the Normal distribution and may be unreliable if the site characteristics are incompatible with the Normal distribution. For example, a site characteristic may take a few discrete values or the regions may occupy areas of site-characteristic space that are not approximately ellipsoidal. Discriminant analysis is not invariant to nonlinear transformations of the site characteristics, and the need for transformation may not be apparent from casual inspection of the data. It is also more difficult than would be the case with a simpler regionalization procedure, such as the index-flood procedure, to estimate the accuracy of the final quantile estimates.

Region of influence

In a fractional-membership procedure the explicit construction of regions is not necessary, as noted by Acreman and Wiltshire (1989). Instead, each site may be regarded as a one-site region. Estimation of parameters or quantiles for any particular site, the "site of interest," can be based on a regional frequency analysis in which a region is chosen to consist of sites that are expected to have a similar frequency distribution to that of the site of interest. These sites constitute what Burn (1990) terms the *region of influence* for the site of interest.

If a set $\{z_1^{(i)}, \ldots, z_m^{(i)}\}$ of site characteristics is available at site i, the separation of the two sets of site characteristics may be measured by the weighted Euclidean distance

$$d(i, j) = \left\{ \sum_{k=1}^{m} w_k (z_k^{(i)} - z_k^{(j)})^2 \right\}^{1/2} \qquad (8.1)$$

From these distances a set of weights $\eta_{i,j}$ is constructed: $\eta_{i,j}$ is the relative weight given to site j in the estimation of the frequency distribution at site i. The sites j for which $\eta_{i,j} > 0$ constitute the region of influence for site i. For given i, $\eta_{i,j}$ should be a decreasing function of $d(i, j)$. It is not necessary to have $\eta_{i,j} = \eta_{j,i}$. If a parameter or quantile θ is estimated from the site-j data by $\hat{\theta}^{(j)}$, then the region-of-influence estimator of θ for site i is

$$\hat{\theta}^{(i)}_{\mathrm{ROI}} = \frac{\sum_j \eta_{i,j} n_j \hat{\theta}^{(j)}}{\sum_j \eta_{i,j} n_j}, \tag{8.2}$$

where n_j is the record length at site j and the sum extends over all sites j that lie within the region of influence of site i.

The region of influence procedure is also attractive when a smooth transition between regions is required. It avoids one problem occurring when sites are assigned to disjoint regions using any of the methods described in Section 4.1, which is that sites adjacent in site-characteristic space may have very different estimated regional growth curves if they are assigned to different regions.

However, the region of influence procedure contains three sources of ambiguity. First, appropriate site characteristics must be defined; like discriminant analysis, the procedure is not invariant to nonlinear transformations of the site characteristics. Second, the weights w_k in Eq. (8.1) must be defined. Burn (1990) proposed that the weight w_k be based on the importance of the site characteristic z_k as an explanatory variable of the between-site variation of the at-site estimate of the 100-year event, $Q(0.99)$. Cavadias (1990) sketched a similar procedure in which the weights might be based on canonical correlations between site characteristics and at-site quantile estimates. These methods depend on at-site estimates of extreme quantiles, which cannot be reliably estimated (else regional analysis would not be needed in the first place), so it seems unlikely that they can give dependable results. A more promising approach would be to use Burn's method but with L-CV rather than $Q(0.99)$ as the quantity whose between-site variation is to be explained by the site characteristics. This is because L-CV can be more reliably estimated than $Q(0.99)$ from at-site data, and its between-site variation within a region is closely related to the accuracy of regional quantile estimates, as noted in Section 7.2. This problem is analogous to that of deciding appropriate weights to assign to the variables used in a cluster analysis, as noted in Section 4.1; however, in the circumstances described in Chapter 4, neither the assessment of homogeneity nor the estimation of quantiles requires the specification of weights. Third, the weights $\eta_{i,j}$ in Eq. (8.2) must be defined. A balance must be struck between including so few sites in the region of influence that the final estimates have as much variability as at-site estimates, and including so many sites that bias is induced in the final estimate because some sites in the region of influence have frequency distributions markedly different from that of the site of interest. Some kind of heterogeneity test, analogous to that described in Section 4.3, would assist in this decision. Zrinji and Burn (1994) used a test of Chowdhury et al. (1991) for this purpose.

Mapping

When the parameters or quantiles arising from regional frequency analysis are found to vary smoothly with the typical site characteristics of sites in each region, it is possible to construct a map or graph that can be used to estimate the parameters for a site, given its site characteristics. Schaefer (1990) used this approach to model annual maximum precipitation for sites in Washington State; the CV and skewness of a fitted generalized extreme-value distribution were graphed as functions of at-site mean annual precipitation. A similar approach was used by Fill (1994).

Mapping of regional estimates is effective when a smooth relation to site characteristics can be found. Its major disadvantage is shared with the fractional membership and region-of-influence procedures; it is more difficult than would be the case with a simpler regionalization procedure, such as the index-flood procedure, to estimate the accuracy of the final quantile estimates.

A mapping approach can also be used with at-site estimates (e.g., McKerchar and Pearson, 1990). However, at-site estimates are more variable than regional estimates and the variation may obscure any smooth relation to site characteristics that may exist. Models that express quantile estimates as functions of site characteristics by regression of at-site quantile estimates on site characteristics (e.g., Tasker and Stedinger, 1986) may also be treated in this category, though their connection with the methods discussed in the rest of this monograph is tenuous.

Bulletin 17

The procedure recommended in Bulletin 17 of the U.S. Water Resources Council (1976, 1977, 1981) fits a log-Pearson type III distribution to annual maximum streamflows at a single site, the skewness of the logarithmically transformed distribution being obtained by combining a data-based estimate with a value read from a map. The procedure uses regional information insofar as the mapped values are derived from observed skewness statistics at many sites. The procedure is further discussed in Section 8.3.

8.2 Regional shape estimation

If the mean and dispersion of the frequency distribution are estimated from at-site statistics, and the shape parameters are estimated by averaging the at-site shape measures for the sites in a region, we call the resulting procedure a *regional shape estimation* procedure. It is intermediate between pure at-site estimation and the index-flood procedure. It has been used in simulation experiments by Lettenmaier et al. (1987) and Stedinger and Lu (1995). These authors used it to fit a generalized extreme-value distribution and called it the "GEV-2" or "GEV-2/R" procedure, but it can of course be used to fit any distribution.

Regional shape estimation could claim to be the optimal procedure for a region in which the mean and *L*-CV of the at-site frequency distributions varied from site to site but the higher-order *L*-moment ratios were equal at each site. In most real-world applications this would not be a physically plausible pattern of variation, because if *L*-CV varies from site to site, then it is likely that the higher-order *L*-moment ratios do too. The justification for regional shape estimation is therefore more plausibly based on the accuracy with which *L*-moments of various orders can be estimated. In a heterogeneous region in which the site-to-site variation in *L*-CV is large compared with the sampling variability of the at-site sample *L*-CV but the site-to-site variation in *L*-skewness is small compared with the sampling variability of the at-site sample *L*-skewness, it is reasonable to expect that regional shape estimation would provide estimates of *L*-CV and *L*-skewness that are more accurate than those obtained by either at-site estimation or the index-flood procedure.

The performance of regional shape estimation can be assessed for any specified region using the simulation procedures of Chapter 7. For the representative regions defined in Section 7.4 a comparison of the performance of index-flood and regional shape estimation is given in Table 8.2. In the homogeneous regions R1 and R3, quantile estimates obtained from the index-flood procedure are considerably more accurate than those obtained from regional shape estimation. This is to be expected because the assumptions of the index-flood procedure, more restrictive than those of regional shape estimation, are satisfied by homogeneous regions. In the heterogeneous regions R2 and R4, regional shape estimation has lower absolute bias than the index-flood procedure and gives more accurate estimates of the most extreme quantiles. The advantage of regional shape estimation over the index-flood procedure for estimation of extreme upper-tail quantiles is stronger for Region R4, which has lower *L*-CV than Region R2. This agrees with the simulation results of Stedinger and Lu (1995) but conflicts with those of Lettenmaier et al. (1987); as noted by Stedinger and Lu (1995), the frequency distributions used in the simulations of Lettenmaier et al. (1987) yielded negative data values with higher frequency than would be reasonable in most applications of regional frequency analysis. Stedinger and Lu (1995) also found that the performance of regional shape estimation relative to that of the index-flood procedure improved as the at-site record lengths increased. This is to be expected because larger at-site record length enables greater accuracy in the at-site estimator of *L*-CV used in regional shape estimation.

Table 8.2. *Comparison of index-flood and regional shape estimators.*

Region		Index flood			Regional shape		
	F:	0.9	0.99	0.999	0.9	0.99	0.999
R1	Bias	−0.2	−2.0	−3.9	0.0	−1.6	−3.5
	Abs. bias	0.2	2.0	3.9	0.1	1.6	3.5
	RMSE	9.2	11.0	14.6	12.6	16.6	20.4
R2	Bias	0.1	−1.2	−2.1	0.1	−1.8	−3.5
	Abs. bias	3.7	10.4	16.6	0.3	3.8	8.8
	RMSE	10.2	15.7	23.0	12.7	17.1	22.3
R3	Bias	0.0	−0.1	−0.1	0.0	−0.1	−0.1
	Abs. bias	0.0	0.1	0.1	0.0	0.1	0.1
	RMSE	2.7	3.0	3.7	3.4	4.6	5.7
R4	Bias	0.0	−0.1	−0.1	0.0	−0.2	−0.4
	Abs. bias	1.8	4.1	6.0	0.2	1.4	2.6
	RMSE	3.3	5.3	7.3	3.4	4.9	6.4

Note: Tabulated values are the regional average relative bias, absolute relative bias, and relative RMSE of estimated quantiles, expressed as percentages, that is, $B^{R}(F)$, $A^{R}(F)$, and $R^{R}(F)$ as defined in Eqs. (7.12)–(7.14).

In summary, it appears that regional shape estimation may be preferred to the index-flood procedure if the following conditions are satisfied: there are doubts about the homogeneity of the region; the main interest is in estimation of quantiles in the extreme upper tail of the frequency distribution; the regional average *L*-CV is fairly low; and at-site record lengths are fairly large but not so large that at-site estimation of *L*-skewness is more accurate than regional shape estimation. These conclusions are based on regions, such as Regions R2 and R4, in which there is strong association between the population *L*-CV and *L*-skewness of the at-site frequency distributions; in the absence of such an association, the advantage of regional shape estimation over the index-flood procedure can be considerably less. It is therefore difficult to identify quantitative conditions under which regional shape estimation is advantageous.

8.3 The Bulletin 17 estimation procedure

Bulletin 17 of the U.S. Water Resources Council (1976, 1977, 1981) recommends a frequency analysis procedure for use by U.S. federal agencies. The procedure is widely used in the United States, Australia, and some other countries. It is described here for comparison with regional frequency analysis and the index-flood procedure.

The Bulletin 17 procedure takes the distribution of the quantity of interest Q to be log-Pearson type III. Estimation is by fitting a Pearson type III distribution to observations of $\log_{10} Q$ using the method of moments. The first two moments, mean and standard deviation, are estimated from the at-site data. The skewness is estimated by a weighted average of the sample skewness and a "generalized skew coefficient" that is derived from regional information. In the absence of detailed studies, Plate I of the Bulletin provides a map that depicts the generalized skew coefficient of annual maximum streamflows in the United States, obtained by drawing smooth isolines on a map of the sample skewness values based on data through 1973 at 2,972 gaging stations.

The estimation procedure in its basic form is as follows.

1. Set aside any exact zero values in the data.
2. Transform the data by taking logarithms to base 10.
3. Calculate the (conventional) moments of the transformed data, $\bar{x}$, s, and g.
4. Test for the presence of low outliers. These are values less than $\bar{x} - K_n s$, where K_n is a threshold value dependent on the sample size n and tabulated in the Bulletin. If any low outliers are found, set them aside and recompute the moments.
5. Read the generalized skew coefficient from the map. Call it g_{map}.
6. Calculate a weighted average of the sample skewness and map skewness, the weights being inversely proportional to the estimated mean square errors of the two estimators. The mean square error of g is computed, from Eq. (6) of the Bulletin, as a function of g itself and the sample size. The mean square error of g_{map} is regarded as a fixed value of 0.302.
7. Estimate the parameters of the Pearson type III distribution from the calculated moments.
8. If any zero values or low outliers are present, apply a "conditional probability adjustment" to allow for them. If the quantile function of the Pearson type III distribution estimated at the previous step is $x(.)$ and the proportion of zero values and outliers in the data is p_0, then the "conditional probability adjusted" estimate of the quantile of nonexceedance probability F is $x\left(\frac{F-p_0}{1-p_0}\right)$. For three values of F, these estimates are calculated and are used to calculate "synthetic statistics," new versions of the moments, which are then used in the method of moments to estimate the parameters of the final Pearson type III distribution of $\log_{10} Q$.

The Bulletin 17 procedure is fundamentally an at-site estimation procedure based on conventional moments of logarithmically transformed data, with two modifications: an adjustment for low outliers and the inclusion of the regionally estimated generalized skew coefficient. It can be criticized on several grounds.

Logarithmic transformation can cause low data values to have undue influence on estimated quantiles in both the lower and the upper tail of the frequency distribution, as noted in Chapter 1. The use of an adjustment for low outliers attempts to allow for this, but it could be argued that the outlier adjustment is a complication that would not have been necessary had the logarithmic transformation not been used. The criterion used to test whether an observation is an outlier is in any case arbitrary, being based on an outlier test for samples from the Normal distribution at a subjectively chosen significance level of 10%. The Bulletin gives no justification for this choice, beyond saying that it was based on "comparing results" of several procedures.

The use of a generalized skew coefficient based on a map is a strange choice for streamflow data. One might expect the skewness of the frequency distribution of streamflow to have sharp discontinuities as a function of location. Consider a site downstream of the confluence of two rivers and sites on the two upstream branches. It is plausible that the shape of the frequency distribution could be very different at the three sites. There is certainly no reason to believe that skewness should be a smooth function of location, as implied by the map in Plate I of Bulletin 17.

The use of conventional moments, particularly the skewness, is another questionable aspect of the method. The skewness of $\log_{10} Q$ for U.S. streamflow data is typically between -1.2 and $+1.2$, and near the extremes of this range the bias of the skewness statistic g can be substantial (Wallis et al., 1974). Parameter estimation by the method of moments may also be inadequate, because other estimators based on different moment-like statistics have better efficiency and robustness (Arora and Singh, 1989).

The main defect of the Bulletin 17 procedure, however, is that it is principally an at-site procedure and does not make sufficient use of regional information. When $n \geq 30$ and $|g| \leq 1$, the weighted average of g and g_{map} gives greater weight to the at-site estimate g. Therefore, except at sites with very short records, the Bulletin 17 procedure makes little use of regional information. In practice, this means that quantile estimates at different sites obtained by the Bulletin 17 procedure often differ by amounts too large for physical reasoning to explain.

Comparisons of the Bulletin 17 and index-flood estimation procedures have been made by Wallis and Wood (1985) and Potter and Lettenmaier (1990). In both papers the index-flood procedure used regional average probability weighted moments to fit a generalized extreme-value distribution. Wallis and Wood used artificial data simulated from the log-Pearson type III distribution for a heterogeneous region with 20 sites; Potter and Lettenmaier used real streamflow data from a 40-site region in Wisconsin and an 80-site region in New England and estimated the accuracy of quantile estimates by a resampling procedure similar to the "bootstrap" of Efron (1982). In each case the RMSE of estimates of extreme quantiles in the upper tail

Table 8.3. *Accuracy of quantile estimates from Bulletin 17 and index-flood procedures.*

Source	Region	Method	F	RMSE
WW	Artificial	Regional GEV/PWM	0.998	17.2
		Bulletin 17	0.998	35.5
PL	Wisconsin	Regional GEV/PWM	0.99	19.6
		Bulletin 17	0.99	37.4
PL	New England	Regional GEV/PWM	0.99	18.6
		Bulletin 17	0.99	41.7

Note: Results taken from Wallis and Wood (1985), "WW," and Potter and Lettenmaier (1990), "PL". RMSE is the regional average relative RMSE of estimated quantiles expressed as a percentage, that is, $R^{\mathrm{R}}(F)$ as defined in (7.14).

of the frequency distribution was smaller for the index-flood procedure, by a factor of about 2. Some example results are given in Table 8.3.

Landwehr, Tasker, and Jarrett (1987) managed to construct two regions in which Bulletin 17 outperforms the index-flood procedure, but these regions are excessively heterogeneous. We simulated realizations of their regions, and found that the average value, over 1,000 realizations of each region, of the heterogeneity measure H defined in Eq. (4.5) was 8.7 for Region 1A and 10.4 for Region 1B. Regional frequency analysis should not be used in regions as heterogeneous as this, and when using the methods described in this monograph there is no reason why it need be. Even the region used by Wallis and Wood (1985) yields an average H of 4.5. We conclude that, even when applied to a region that is heterogeneous and misspecified, regional frequency analysis using an index-flood procedure based on probability weighted moments or L-moments is likely to be overwhelmingly superior to the Bulletin 17 procedure.

8.4 Quantile estimation at ungaged sites

Estimating the frequency distribution at a site for which no measurements of the quantity of interest are available is sometimes required. We call such a site an "ungaged site," using, for convenience, terminology common in the analysis of streamflow data. Typically some site characteristics for the ungaged site are known. These must be used to assign the site to a suitable region and to estimate the index flood, usually the mean of the at-site frequency distribution, at the ungaged site.

Assignment of an ungaged site to a region will often not be a problem. If the formation of regions was done in accordance with the principles outlined in Section 4.1, using site characteristics rather than at-site statistics, then the site characteristics available at the ungaged site will often suffice to identify the region to which it should belong. For example, if regions are formed by cluster analysis on site characteristics using Ward's method, then an ungaged site would most naturally be assigned to the region whose center is closest, in the space of site characteristics used in the clustering, to the site characteristics of the ungaged site. Difficulties arise if the site characteristics used to form regions include some that are derived from at-site data measurements, as noted on page 55. In this case, the unobserved site characteristics must be estimated, or a method of assigning sites to regions must be found that involves only the observed site characteristics.

Estimation of the index flood at an ungaged site may require further modeling. In some cases, the index flood may vary slowly enough geographically, or over some space of site characteristics, that it can be mapped and its value for an ungaged site inferred from contours drawn on the map. When the relation between the index flood and site characteristics is not so clear, a formal statistical model may be used. A simple example is a linear regression model relating the index flood μ_i, or some function of it such as $\log \mu_i$, to a linear combination of site characteristics measured at site i, $z_j^{(i)}$, $j = 1, \ldots, k$:

$$\mu_i = \theta_0 + \sum_{j=1}^{k} \theta_j z_j^{(i)} + u_i \,. \tag{8.3}$$

The model parameters θ_j, $j = 0, 1, \ldots, k$, and the variance of the error term u_i can be estimated by fitting the model to sites at which data are available. These sites need not constitute a homogeneous region, because a model such as Eq. (8.3) may well provide an adequate approximation for a wider range of sites.

Models of the form (8.3) have been widely used in hydrology to estimate the mean annual maximum streamflow at ungaged sites. Often the at-site sample mean is substituted for the population mean μ_i in Eq. (8.3) and the model is fitted by the method of least squares. This may give adequate estimates of the θ_j parameters but can seriously overestimate the error of estimation of μ_i. To see this, write the fitted model as

$$\bar{x}_i = \theta_0 + \sum_{j=1}^{k} \theta_j z_j^{(i)} + v_i \,, \tag{8.4}$$

$$v_i = u_i + (\bar{x}_i - \mu_i) \,. \tag{8.5}$$

The error term v_i in Eq. (8.4) contains both u_i, the error in approximating μ_i by $\theta_0 + \sum_j \theta_j z_j^{(i)}$, and $\bar{x}_i - \mu_i$, the sampling error of the at-site mean $\bar{x}_i$. The term $\bar{x}_i - \mu_i$ has variance σ_i^2 / n_i, where σ_i^2 is the variance of the frequency distribution at site i and n_i is the record length at site i; this may be considerably larger than the variance of u_i. Thus the residual variance from fitting model (8.4), an estimate of the variance of v_i, includes a contribution from the sampling error of the at-site sample means that causes it to be an overestimate of the variance of u_i, the true error associated with model (8.3).

A more careful approach to the estimation of model (8.3) seeks to separate the contributions of model error and sampling error. Stedinger and Tasker (1985, 1986) have described such an approach, based on generalized least-squares fitting of model (8.3). Though the calculations are somewhat involved, the results can be greatly superior to the use of ordinary least-squares regression. We sketch their method for an application to maximum streamflow data in which the index flood is the mean annual maximum streamflow, estimated by the at-site sample mean, and the only site characteristic in the model is the logarithm of the drainage area A of the gaging site.

The basic model for μ_i is

$$\mu_i = \theta_0 + \theta_1 \log A_i + u_i \,, \qquad i = 1, \ldots, N, \tag{8.6}$$

where the model errors u_i at the N sites are assumed to be independent and identically distributed with mean zero and variance ω^2. In terms of the observable quantity $\bar{x}_i$ the model becomes

$$\bar{x}_i = \theta_0 + \theta_1 \log A_i + u_i + e_i \,, \tag{8.7}$$

where the sampling errors $e_i = \bar{x}_i - \mu_i$ have mean zero and covariances

$$\mathrm{E}(e_i e_j) = \Sigma_{ij} = \rho_{ij}\sigma_i\sigma_j n_{ij}/(n_i n_j)\,. \tag{8.8}$$

Here σ_i is the standard deviation of the frequency distribution at site i, ρ_{ij} is the correlation between the frequency distributions at sites i and j, n_i is the record length at site i, and n_{ij} is the number of concurrent observations at sites i and j. The quantities ρ_{ij} and σ_i are unknown and must be estimated, but for the moment we assume them to be known. Model (8.7) can be written in vector form as

$$\mathbf{x} = \mathbf{Z}\boldsymbol{\theta} + \mathbf{v} \tag{8.9}$$

where $\mathbf{x}$ is an N-vector with ith element $\bar{x}_i$, $\mathbf{Z}$ is an $N\times 2$ matrix whose first column has each element equal to 1 and whose second column has ith element $\log A_i$, $\boldsymbol{\theta} = [\theta_0 \ \theta_1]^{\mathrm{T}}$, and $\mathbf{v}$ is an N-vector with mean zero and covariance matrix $\mathbf{G} = \omega^2\mathbf{I}+\boldsymbol{\Sigma}$; here $\mathbf{I}$ is the $N\times N$ identity matrix and $\boldsymbol{\Sigma}$ is an $N\times N$ matrix with (i, j)

element Σ_{ij}. To estimate the unknown parameters $\boldsymbol{\theta}$ and ω^2, Stedinger and Tasker (1985) compared ordinary, weighted, and generalized least-squares methods. The best results were obtained using the generalized least-squares method. If ω^2 were known, the minimum-variance unbiased estimator of $\boldsymbol{\theta}$ would be the generalized least-squares estimator

$$\hat{\theta} = (\mathbf{Z}^{\mathrm{T}}\mathbf{G}^{-1}\mathbf{Z})^{-1}\mathbf{Z}^{\mathrm{T}}\mathbf{G}^{-1}\mathbf{x} \tag{8.10}$$

and would satisfy

$$\mathrm{E}\{(\mathbf{x} - \mathbf{Z}\hat{\boldsymbol{\theta}})^{\mathrm{T}}\mathbf{G}^{-1}(\mathbf{x} - \mathbf{Z}\hat{\boldsymbol{\theta}})\} = N - 2\,. \tag{8.11}$$

In practice ω^2 is unknown. It can be estimated by an iterative procedure in which a trial value of ω^2 is chosen, $\mathbf{G}$ and thence $\hat{\boldsymbol{\theta}}$ in Eq. (8.10) are calculated, and the quantity

$$(\mathbf{x} - \mathbf{Z}\hat{\boldsymbol{\theta}})^{\mathrm{T}}\mathbf{G}^{-1}(\mathbf{x} - \mathbf{Z}\hat{\boldsymbol{\theta}}) - (N - 2) \tag{8.12}$$

is compared with its expected value of zero. The value of ω^2 for which Eq. (8.12) is zero is taken as the estimate of ω^2, and the corresponding value of $\hat{\boldsymbol{\theta}}$ is the estimate of $\boldsymbol{\theta}$.

There remains the question of estimation of ρ_{ij} and σ_i. Assuming that intersite dependence does not vary too much from site to site, Stedinger and Tasker (1985) suggest that each ρ_{ij} can be estimated by the average intersite correlation calculated from the various sites' data. This may well be adequate in practice and should be better than ignoring intersite dependence altogether. More elaborate estimators, obtained for example by relating ρ_{ij} to site characteristics using another regression model, might be considered. For estimation of σ_i, Stedinger and Tasker (1985) found that the simple substitution of the at-site sample standard deviation was not adequate. They preferred to estimate σ_i from another regression model, relating σ_i to $\log A_i$ with a multiplicative error:

$$\sigma_i = (\phi_0 + \phi_1 \log A_i)\exp(d_i), \tag{8.13}$$

where the errors d_i are independent and Normally distributed with mean $-\delta^2/2$ and variance δ^2; thus $\mathrm{E}\{\exp(d_i)\} = 1$. Estimation of model (8.13) involves a simplified version of the generalized least-squares procedure used for model (8.6); in particular, intersite dependence is ignored (Stedinger and Tasker, 1985, Appendix B).

8.5 Measurement error

In frequency analysis it is normally assumed that the data values provide a true representation of the underlying values of the quantity of interest. However, in many applications data are obtained by a physical measurement process that may be subject to error. Measurement error, if large enough, can cause bias and loss of accuracy in frequency analysis. The seriousness of the effect, and the kind and magnitude of errors that may arise, are very dependent on the type of data and the measurement process. Subject-matter knowledge of the field of application is essential for an accurate assessment of the significance of measurement error in any particular analysis.

For the analysis of streamflow data, Potter and Walker (1981, 1985) drew attention to measurement error and showed that it might lead to overestimation of the moments of the frequency distribution. Cong and Xu (1987) and Kuczera (1992) investigated the consequences for at-site frequency analysis and found that the effect of measurement error on quantile estimates can sometimes be large enough to be of practical concern, particularly when the coefficient of variation of the frequency distribution is low.

To illustrate the effects of measurement error, we use a simple model similar to that of Potter and Walker (1981) and apply it to two of the representative regions of Section 7.4. Suppose that the actual values of the quantity of interest at site i are Q_{ij}, $j = 1, \ldots, n_i$, but that the observed values include a multiplicative error that has a lognormal distribution with mean 1 and variance σ_e^2. Thus an observed value would be

$$\tilde{Q}_{ij} = Q_{ij} e_{ij}, \tag{8.14}$$

where $\log e_{ij}$ is Normally distributed with mean $-\frac{1}{2}\log(1+\sigma_e^2)$ and variance $\log(1+\sigma_e^2)$. We assume that the errors e_{ij} are independent for all i and j. The actual values Q_{ij} are generated from the frequency distributions of Regions R1 and R2 defined in Section 7.4. These 15-site regions, one homogeneous and one heterogeneous, have generalized extreme-value frequency distributions with regional average L-moment ratios $\tau = 0.25$ and $\tau_3 = 0.25$. The standard deviation of the measurement error, σ_e, is set to 0, 0.1 or 0.2, which we refer to as "no error", "10% error", and "20% error", respectively.

The distribution of the observed values $\tilde{Q}_{ij}$ is slightly more skew than that of the true values Q_{ij}. This is illustrated in Figure 8.1 for the case in which Q_{ij} has the generalized extreme-value distribution of Region R1 and $\tilde{Q}_{ij}$ includes 20% measurement error. The distribution of $\tilde{Q}_{ij}$ has $\tau = 0.267$ and $\tau_3 = 0.256$, compared with $\tau = 0.25$ and $\tau_3 = 0.25$ for the distribution of Q_{ij}. As with the

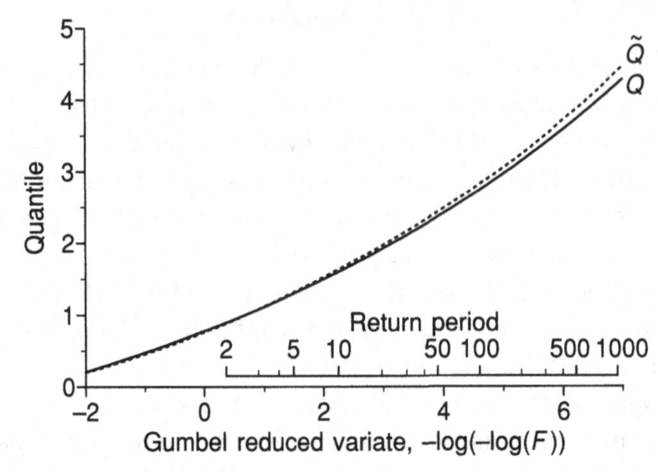

Fig. 8.1. Quantile functions for the sites in Region R1: true values Q_{ij} (solid line), and observed values $\tilde{Q}_{ij}$ including 20% measurement error (dotted line).

error model of Potter and Walker (1981), the inclusion of measurement error tends to inflate the *L*-CV and *L*-skewness of the observed data and may be expected to lead to overestimation of the upper-tail quantiles of the frequency distribution.

The performance of regional frequency analysis in the presence of measurement error was evaluated by simulation, using the procedure described in Section 7.3, with data generated according to the measurement-error model (8.14). The generalized extreme-value distribution was fitted by the regional *L*-moment algorithm. Table 8.4 contains a summary of the results for estimation of the quantiles and growth curve. Results for the case of no error are of course identical to those in Table 7.2. The addition of measurement error causes the bias to become increasingly positive and tends to increase the variability of quantile estimates. The effect is barely perceptible with 10% measurement error; indeed, in some cases the negative bias present when there is no measurement error is reduced when the error is 10%, and the resulting estimates have lower RMSE. With 20% error, however, the estimates are clearly less accurate than in the absence of measurement error. The effect is more serious for the homogeneous region; for example, in Region R1 the RMSE of the quantile estimate $\hat{Q}(0.99)$ is increased by 23% relative to the case of no measurement error, whereas in Region R2 the corresponding increase is 15%.

In this example, typical measurement errors would need to be close to 20% before a significant effect on the accuracy of quantile estimates would be felt. We emphasize, however, that this is purely an illustrative example; different regions or different patterns of measurement error could lead to very different results.

Table 8.4. *Simulation results for regions with data subject to measurement error.*

Region	Error		Quantiles			Growth curve		
		F:	0.9	0.99	0.999	0.9	0.99	0.999
		Bias	−0.2	−2.0	−3.9	−0.2	−2.0	−3.9
R1	0	Abs. bias	0.2	2.0	3.9	0.2	2.0	3.9
		RMSE	9.2	11.0	14.6	1.3	5.7	11.0
		Bias	0.7	−0.2	−1.3	0.7	−0.2	−1.4
R1	10%	Abs. bias	0.7	0.2	1.3	0.7	0.2	1.4
		RMSE	9.5	11.2	14.7	1.5	5.5	10.7
		Bias	3.0	5.2	6.6	3.0	5.1	6.5
R1	20%	Abs. bias	3.0	5.2	6.6	3.0	5.1	6.5
		RMSE	10.8	13.5	17.8	3.3	7.8	13.5
		Bias	0.1	−1.2	−2.1	0.1	−1.2	−2.2
R2	0	Abs. bias	3.7	10.4	16.6	3.7	10.4	16.6
		RMSE	10.2	15.7	23.0	4.1	12.2	20.6
		Bias	1.0	0.7	0.5	1.0	0.6	0.4
R2	10%	Abs. bias	3.8	10.4	16.9	3.8	10.4	16.8
		RMSE	10.6	16.0	23.5	4.1	12.3	21.0
		Bias	3.4	6.1	8.7	3.3	6.1	8.6
R2	20%	Abs. bias	4.5	11.6	18.8	4.5	11.6	18.8
		RMSE	11.9	18.0	26.4	4.8	13.6	23.3

Note: Tabulated values are the regional average relative bias, absolute relative bias, and relative RMSE of estimated quantiles, expressed as percentages, that is, $B^R(F)$, $A^R(F)$, and $R^R(F)$ as defined in Eqs. (7.12)–(7.14), and the corresponding quantities for the estimated growth curve. "Error" is the standard deviation of the measurement error, σ_e, expressed as a percentage.

8.6 Historical information

In addition to a sequence of measurements of the quantity of interest, there may in some circumstances be additional information about the magnitudes of events that occurred outside the period of systematic measurement. For example, with annual maximum streamflow data, in addition to the gaged record there may be information about large flood events that occurred before the period of continuous gaging. Such information may be historical, in the form of recollections or records left by human observers, or paleological, based on botanical or geophysical evidence.

Some difficulties are associated with historical information that do not arise with the systematic gaged record.

First is the question of deciding exactly what information is present at each site. For example, consider a set of annual maximum streamflow data (site 01626000 in the example described in Section 9.2) with a gaged record extending from 1953 to 1991, the measurements ranging from 419 $\mathrm{ft}^3\mathrm{s}^{-1}$ to 17,500 $\mathrm{ft}^3\mathrm{s}^{-1}$, and a historic event of 14,500 $\mathrm{ft}^3\mathrm{s}^{-1}$ recorded in 1943 that was exceeded twice in the gaged record. A complete specification of the historical information requires us to identify the possible maximum streamflow values in all years prior to 1953. How large a flood, for example, could have occurred in any of the years 1944–52, or in the years up to 1942, without some information about it having been recorded? It is plausible that there exists a "threshold of perception" such that only floods that exceeded this threshold would leave records for posterity (Gerard and Karpuk, 1979). Accurate specification of historical information requires that this threshold be accurately estimated throughout a period at least as far back as the earliest historical event and probably for some time before that.

A second concern is the consistency of historical information across different sites. The effect on quantile estimates of heterogeneity in a region is likely to be exacerbated if historical information is concentrated at a few sites that are not typical of the region as a whole. Historical streamflow information, for example, is most easily obtained for large or densely populated drainage basins, whose frequency distributions tend to have relatively low *L*-CV and *L*-skewness. If the results are not to be biased, historical information should be developed for all sites and years back to the earliest recorded year of historical information. In practice this is rarely done and historical information is acquired in a haphazard and inconsistent manner.

A third concern is the accuracy of historical data. The methods used to obtain historical and paleological streamflow data tend to be less accurate than direct gaging of the flow. Historical events, because they tend to be the most extreme, are often the most difficult to measure accurately. Thus measurement error can be a more significant problem when historical information is used than for the gaged record alone.

Finally, historical and paleological information may date from so far in the past that the frequency distribution has changed in the intervening period. In some environmental applications this can limit the utility of information about events more than 100 or 200 years in the past. Paleoflood data for the southwestern United States have been used both to fit a single frequency distribution applicable throughout the last 2,000 years (Kochel et al., 1982) and to identify changes in the climate during this period (Ely et al., 1993); one must regard these applications as incompatible.

The incorporation of historical information into frequency analysis using *L*-moments is possible for some specifications of historical data. Suppose that the data available at a site are h "historic" values, assumed to be the h largest in a period of m years, and n recent values from the period of systematic measurement. Some of the historic values may be from the systematic record too, so the actual number of systematic observations may be greater than n. In the absence of more detailed information, it is reasonable to suppose that the recent values are a random sample, drawn without replacement, of the $m - h$ smallest values in the total period of m years. Ding and Yang (1988) and Wang (1990b) have derived estimators of the probability weighted moments of the frequency distribution for this type of historical data. The use of probability weighted moments and L-moments in related censored-data problems is discussed by Wang (1990a) and Hosking (1995). The adaptation of these estimators to regional frequency analysis remains problematical; in the index-flood procedure, an appropriate way of choosing the weights in the weighted average (1.5) is difficult to derive. L-moment methods are even less suitable for dealing with some other specifications of historical information, such as when the available data are the number of times in a given period that a certain threshold was exceeded.

For regional frequency analysis using historical information, the method of Jin and Stedinger (1989) appears to be the best currently available. It fits a regional generalized extreme-value distribution and combines features of maximum-likelihood estimation, which has great flexibility in dealing with different specifications of historical information, with the method of L-moments, which gives accurate estimates of the parameters of the generalized extreme-value distribution.

If accurately specified historical information is available, the accuracy with which quantiles of the frequency distribution can be estimated can be greatly increased. This has been documented for single-site frequency analysis by several authors (e.g., Leese, 1973; Tasker and Thomas, 1978; Cohn and Stedinger, 1987) and by Jin and Stedinger (1989) for their regional frequency analysis procedure. However, it is not clear that these ideal conditions can often be attained in practice, because of the difficulties noted above. In particular, the greater measurement error associated with historical data has only rarely been taken into account in published assessments of the value of historical information (Cong and Xu, 1987; Kuczera, 1992). We therefore remain somewhat skeptical about the practical utility of historical information.

9

Examples

9.1 U.S. annual precipitation totals

9.1.1 Introduction

In 1989 the U.S. Army Corps of Engineers was charged with the responsibility of conducting a national study of water management during periods of drought. One of the results of the study is the National Drought Atlas (Willeke et al., 1995), which contains analyses of data on monthly precipitation, streamflow, reservoir levels, and the Palmer Drought Index for over 1,000 measuring sites in the continental United States. Analysis of the precipitation data used regional frequency analysis and was based on *L*-moments. Precipitation data were available as totals, in inches, for durations of 1, 2, 3, 6, 12, 24, 36 and 60 months starting in each calendar month January through December. Though regions could in principle have been defined separately for each combination of duration and starting month, this would have led to an atlas that would have been excessively large and difficult to use. It was therefore decided to construct a single set of regions, based on the data for annual precipitation totals, and to use these regions when fitting regional frequency distributions to the data for all durations and starting months.

Here we describe the analysis of the data for annual precipitation totals (though data for other durations and starting months affect some parts of the analysis). The analysis illustrates the steps involved in a large-scale regional frequency analysis exercise and shows how some of the commonly occurring problems in regional frequency analysis may be overcome.

Some of the results in this section can also be found in Guttman (1993) and Guttman et al. (1993). Detailed tabulations of the final results – quantile estimates of annual and monthly precipitation for each measuring site – can be found in the National Drought Atlas (Willeke et al., 1995).

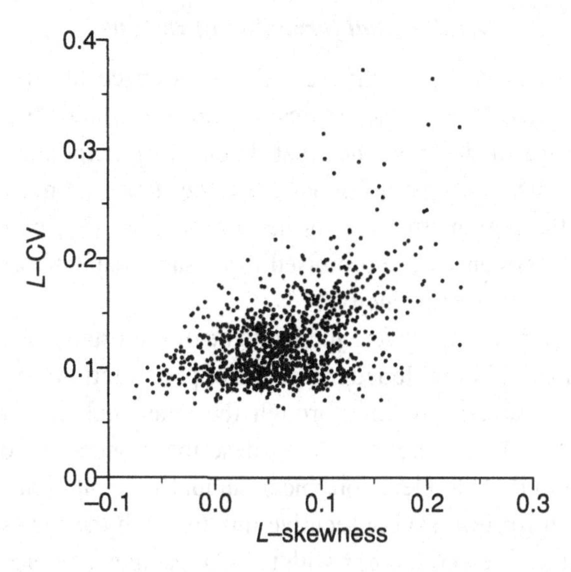

Fig. 9.1. *L*-moment ratios for the U.S. annual precipitation data.

9.1.2 Data

The data set used for the analysis was taken from the Historical Climatology Network (Karl et al., 1990). The Historical Climatology Network is a database containing monthly temperature and precipitation data through 1989 for 1,219 sites in the United States. It was prepared for the U.S. Department of Energy by the National Climatic Data Center (NCDC). The stations in the network are considered to be the best long-term records available. The analysis used data from sites that satisfied the following criteria: record length at least 60 years, not more than 10% of the monthly values missing, and no more than 12 consecutive months of data missing. There were 1,123 such sites, with an average record length of 85 years.

The discordancy statistic defined in Chapter 3 was calculated for the entire set of annual precipitation data, but for a large heterogeneous data set such as this it is not particularly informative. The data had already undergone quality checks at NCDC, and there was no reason to expect that gross errors would be present. The *L*-CV and *L*-skewness of the data are shown in Figure 9.1. When applied to the entire data set, some extremely high values, as high as 17, of the discordancy statistic D_i were observed. These are for sites in the arid southwest of the United States, for which the *L*-CV of annual precipitation totals is in the range 0.3–0.4. These sites are consistent among themselves: that they are discordant with the rest of the data merely indicates that the frequency distributions at the 1,123 sites are very heterogeneous.

9.1.3 Initial formation of regions

Regions were formed by identifying clusters in a space of site characteristics. Clustering was based on site characteristics only and did not involve at-site statistics measuring the shape of the frequency distribution of precipitation. This is in accordance with the discussion in Section 4.1. At-site statistics are used to assess the homogeneity of the regions that are formed in the clustering procedure, and the validity of this assessment is compromised if the same data are used both to form regions and to test their homogeneity.

The site characteristics used were judged to be of importance in defining a site's precipitation climate. They included indicators of precipitation amounts, indicators of the distribution of the amounts through the year, and the site's geographic location. Seven variables were chosen to describe a precipitation climate: site latitude, site longitude, site elevation, mean annual precipitation, the ratio of the mean precipitation for the two consecutive months with the lowest mean amount in the year to that for the two months with the highest mean amount, the beginning month of the two consecutive months with the highest mean amount in the year, and the beginning month of the two consecutive months with the lowest mean amount in the year.

The observed scales of the variables are very different, and standard methods of cluster analysis are very sensitive to such scale differences. The variables were therefore transformed so that their ranges were comparable. The location, precipitation amount, and precipitation ratio variables were rescaled so that their values lay between 0 and 1. The other two variables represent a point along an annual cycle and were transformed by representing the months by a sine curve with a period of one year; the range of the transformed variables is from -1 to $+1$. Table 9.1 shows the transformations from the seven site characteristics to the variables used in cluster analysis.

There is some arbitrariness involved in the choice of transformations. A more detailed analysis might seek to identify site characteristics that are likely to be particularly important in defining a precipitation climate, and transform these characteristics so that their range is greater than that of the less important characteristics. The variables representing a point in an annual cycle could be replaced by two transformed variables, $\cos(2\pi X/12)$ and $\sin(2\pi X/12)$, that correspond to the coordinates of a point on the unit circle. This would avoid irregularities in the present transformation, an example being that the cases $X = 6$ and $X = 12$ yield the same Y value. However, a successful clustering was attained using the actual Y variables of Table 9.1.

Cluster analysis was performed using SAS average linkage and Ward's minimum variance hierarchical clustering software (SAS, 1988). In the average-linkage

Table 9.1. *Transformation of site characteristics.*

Site characteristic, X	Cluster variable, Y
Latitude (deg)	$Y = X/90$
Longitude (deg)	$Y = X/150$
Elevation (ft)	$Y = X/10000$
Mean annual precipitation (in)	$Y = X/100$
Ratio of minimum average two-month precipitation to maximum average two-month precipitation	$Y = X$
Beginning month of minimum average two-month precipitation (Jan.=1, . . . , Dec.=12)	$Y = \sin(2\pi X/12)$
Beginning month of maximum average two-month precipitation (Jan.=1, . . . , Dec.=12)	$Y = \sin(2\pi X/12)$

method the distance between two clusters is the average Euclidean distance between two observations, one in each cluster. Clusters with small variance tend to be joined, and the procedure is biased in favor of producing clusters with equal dispersion in the space of clustering variables. In Ward's method, the distance between two clusters is the sum of squares between the two clusters summed over all the variables. The method tends to join clusters that contain a small number of sites and is strongly biased in favor of producing clusters containing approximately equal numbers of sites.

Both clustering methods are based on Euclidean distances and are sensitive to redundant information that may be contained in the variables as well as to the scale of the variables being clustered (Fovell and Fovell, 1993). The four variables describing precipitation characteristics are intended to contain mutually independent information. The location variables contain information that to some extent overlaps the information contained in the other four variables. They were used, however, as proxies for other unmeasured factors that vary smoothly with location.

The output from the average-linkage and Ward's methods was very similar. Clusterings in which the 1,123 sites were divided into between 8 and 60 clusters were initially obtained. The clusters were reviewed to assess whether they were spatially continuous and physically reasonable. A clustering containing about 40 clusters was selected as the basis for further progress. Its clusters were subjectively judged to be reasonable in that the areas that they covered could easily be justified on the basis of the physical processes that control precipitation.

The discordancy and heterogeneity measures D_i and H defined in Eqs. (3.3) and (4.5) were computed for each region identified by the clustering procedure. When the computed heterogeneity measure H exceeded 2, indicating that a region was "definitely heterogeneous", the sites in the region were separated by the

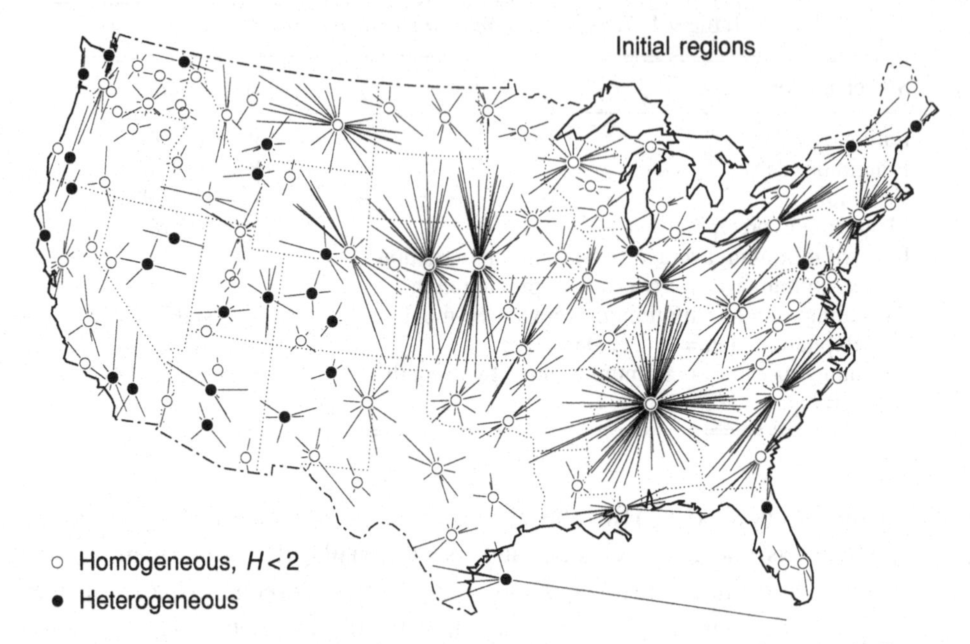

Fig. 9.2. Precipitation regions, initial version. The symbol at the center of a region indicates whether the region is homogeneous. Lines radiating from the center end at the locations of stations in the region.

clustering algorithms into smaller groups. The discordancy measure occasionally indicated that several neighboring sites in a region were discordant with the rest of the region. In these cases, a new region was formed containing just the discordant sites.

These procedures for subdivision of heterogeneous regions continued until no further progress could be made. At this point there were 109 regions, 82 of which were "acceptably homogeneous" or only "possibly heterogeneous", with $H < 2$. The 27 "definitely heterogeneous" regions were mostly in the western states. The regions are shown in Figure 9.2.

9.1.4 Refinement of regions

The regions obtained from the automatic clustering procedure were adjusted manually. Inspection of the clusters, taking into account the topography and spatial patterns of mean annual precipitation in the areas covered by the clusters, suggested several natural and physically reasonable modifications to the clusters, which resulted in more nearly homogeneous clusters; four sites were deleted from the data set during this process.

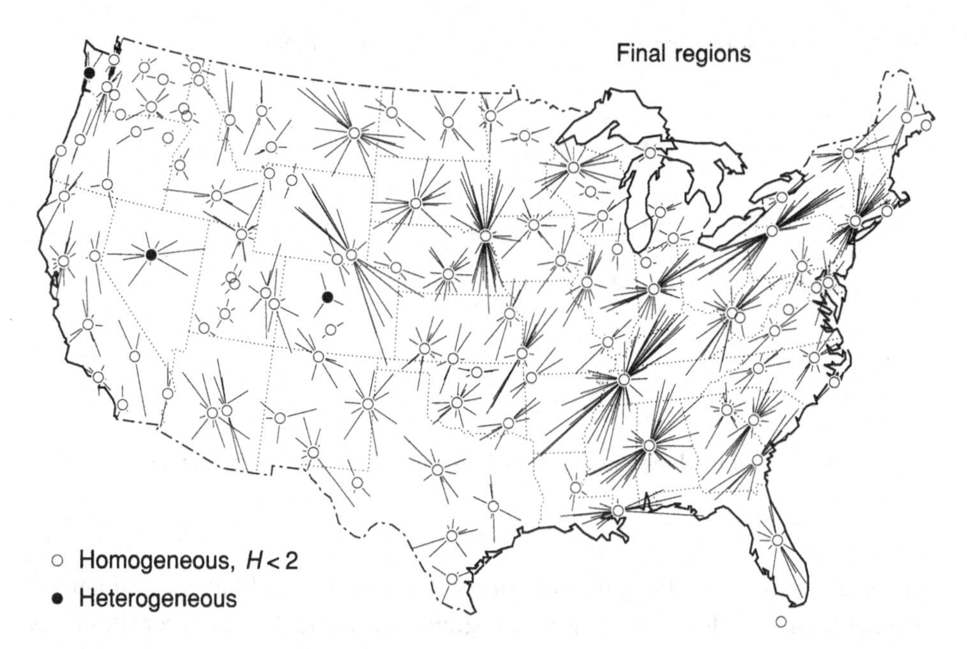

Fig. 9.3. Precipitation regions, final version. The symbol at the center of a region indicates whether the region is homogeneous. Lines radiating from the center end at the locations of stations in the region.

The final set of regions is illustrated in Figure 9.3. The remaining 1,119 sites were grouped into 111 regions. Of these regions, 108 were loosely categorized as homogeneous, with heterogeneity measures $H < 2$; of these 108 regions, 73 were "acceptably homogeneous", with $H < 1$, and 35 were "possibly heterogeneous", with $1 \le H < 2$. Only 3 regions remained "definitely heterogeneous", with $H \ge 2$. The regions contain between 1 and 48 sites; the median region size is 8 sites and 645 station-years of data; half of the sites are in regions containing at least 13 sites.

The modifications to the regions used all of the techniques listed on page 59. As an example, we describe the adjustments made to regions in Utah and Colorado. The regions are mapped in Figure 9.4, which shows both the initial and the final regions. The adjustments were made manually, taking into account the site characteristics, particularly mean annual precipitation and elevation. Figure 9.5 shows the initial regions, designated by letters A through H, together with the mean annual precipitation, in inches, for each of the 44 sites in the eight regions. Three separate adjustments were made.

First, there were several sites that appeared to be out of place in their assigned region and to fit better in a neighboring region. These sites were moved into the regions where they appeared to belong, as indicated by the arrows on Figure 9.5. Four sites were moved into Region C, one each from regions B, D, E, and G. Each

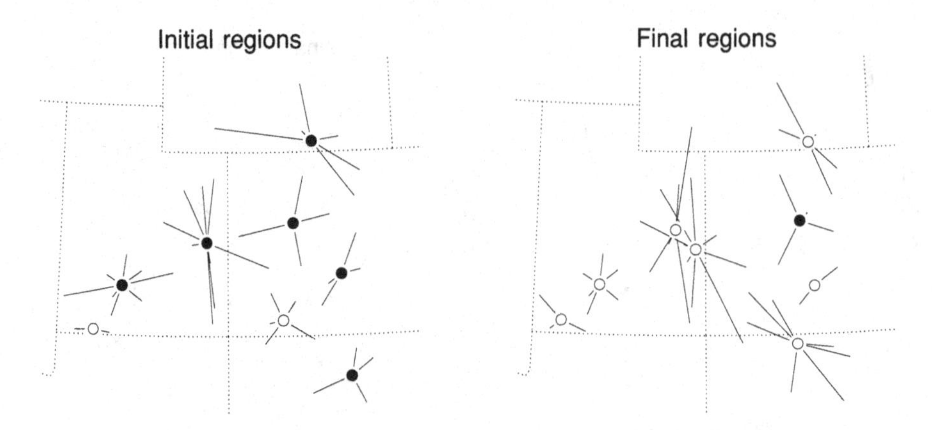

Fig. 9.4. Initial and final regions for annual precipitation totals in Utah and Colorado.

of the sites moved is at the periphery of its initial region and has a mean annual precipitation that is closer to the sites in Region C than to that of the other sites in its initial region. Data on site elevation and the general orography of the area confirmed the physical reasonableness of these moves. On similar grounds, one site was moved from Region B into Region A and one site from Region F into Region E. As a result of this adjustment, Regions B, D, and F became homogeneous, as indicated by a value $H < 2$ for the heterogeneity measure (4.5); Regions C, E, and H remained heterogeneous and Regions A and G remained homogeneous.

Next, Regions G and H were combined. They each contain sites in the river valleys surrounding the San Juan and Sangre de Cristo ranges of the Rocky Mountains in Colorado and New Mexico, and should have similar precipitation climates, because the elevations and mean annual precipitations of the sites in the two regions overlap. The combined region is homogeneous ($H = 1.5$).

Region C, now augmented by four additional sites, was still heterogeneous ($H = 3.1$). In this mountainous area of eastern Utah, elevation and mean annual precipitation are correlated with each other and either one is likely to have a large influence on the shape of the distribution of precipitation totals. When the region was divided into two subregions of roughly equal size, according to the site values of elevation or mean annual precipitation, the subregions were found to be homogeneous. The final choice was a division based on the criterion of whether the site's mean annual precipitation was greater than 8 in. Both subregions were homogeneous ($H = 1.3$ and $H = 1.2$).

The only remaining heterogeneous region was Region E ($H = 2.9$). Not much can be done about this, because one of its sites has a much lower L-CV of its annual precipitation totals than any neighboring site and will cause any region to which it is assigned to be heterogeneous.

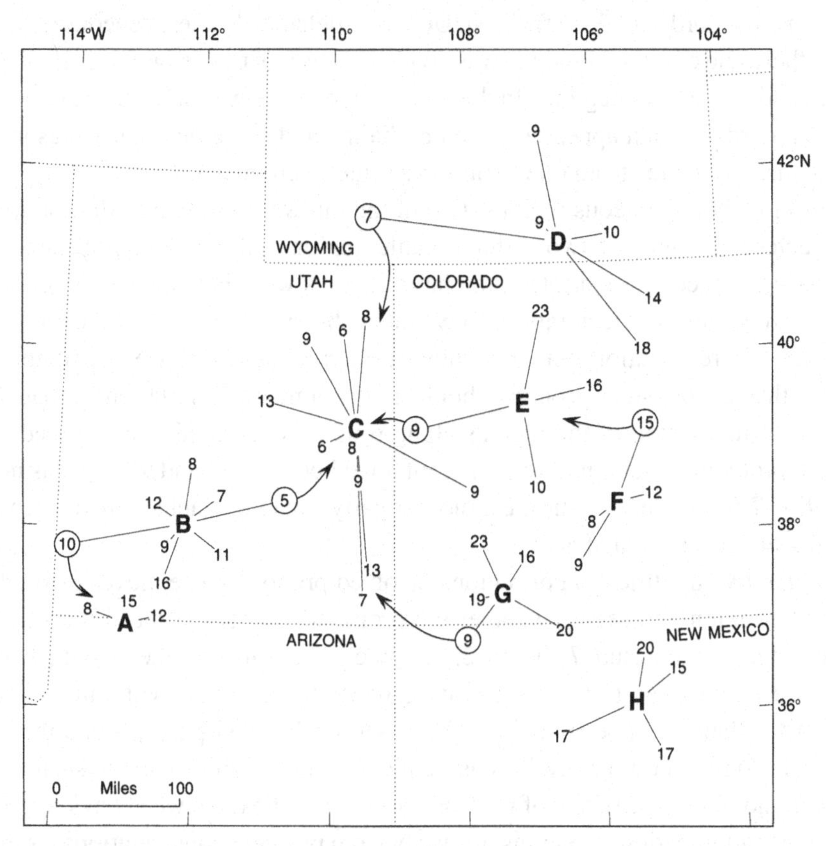

Fig. 9.5. Modification of the regions for annual precipitation totals in Utah and Colorado. Letters A through H identify the regions. Lines join the region center to each site in the region. Small numbers are mean annual precipitation, in inches, for each site. Curved arrows indicate sites that were moved from one region to another.

Similar procedures were followed to modify heterogeneous regions in other parts of the United States. In the northwestern states, a few physically reasonable moves of sites from one region to another achieved homogeneity in all regions except one. In Nevada, it proved impossible to define homogeneous regions on the basis of the available site characteristics; the sites in this state were left as a single very heterogeneous region ($H = 7.3$; at-site sample L-CV of annual precipitation totals varies from 0.15 to 0.28). Sites in southern California were reassigned to three regions suggested by the local topography: inland desert sites and sites near the coast north and south of Los Angeles. Sites in four regions in Arizona were regrouped into two regions according to their mean annual precipitation. The site at Key West, Florida, initially grouped with sites in Texas, did not fit well there nor with regions in Florida. It was split off as a single-site region. In Florida, three regions were combined into one and three sites whose data appeared unreliable, according to records obtained from NCDC, were deleted. A heterogeneous region at the southern end of Lake Michigan was divided according to a geographical split into eastern and western subregions. In the northeast, one site was deleted because from NCDC records its data appeared to be unreliable, and some physically reasonable reassignments of sites from one region to another were made.

Four large homogeneous regions, two in the midwest and two in the southeast, were each subdivided into two or three smaller regions. Although homogeneous for calendar-year precipitation totals, the large regions were heterogeneous for some other starting months and durations. It was also observed that maps of the estimated quantiles of precipitation amounts showed sharp changes between neighboring regions that on physical grounds should have similar precipitation climates. A subjective subdivision of the regions along roughly geographic lines solved both of these problems. The initial regions contained between 37 and 97 sites; as noted in Section 7.6, smaller regions are almost equally capable of achieving the benefits of regional frequency analysis.

The iterative modification of regions involved proposing site moves or subdivisions of a region and seeing whether the proposed changes reduced the value of the heterogeneity measure H below 2. To some extent this uses the H statistic both to choose regions and to test their homogeneity, and conflicts with our advice in Section 4.1 that the same statistics not be used for both purposes. It is difficult to avoid this conflict completely. To ensure that the final regions had physical utility and were not merely artifacts of random variation in the data, we always sought to justify the modification of regions primarily from physical considerations, as in the Utah–Colorado examples discussed above. It is particularly important to emphasize that at no stage was a site removed from a region solely because its sample L-moment ratios were inconsistent with those of the other sites in the region; site reassignments were made only if they seemed reasonable on climatological and meteorological grounds.

9.1.5 Choice of distribution

The goodness-of-fit statistic Z^{DIST} defined in Eq. (5.6) was computed for each of the 108 homogeneous regions for each of five distributions: generalized logistic, generalized extreme-value, lognormal, Pearson type III, and generalized Pareto. The number of regions for which each distribution gave an acceptable fit, with $|Z^{\mathrm{DIST}}| \le 1.64$, is shown in Table 9.2. The lognormal and Pearson type III distributions are acceptable most often, in 92 of the homogeneous regions.

Though a distribution could be chosen separately for each region, convenience of use of a large-scale analysis may suggest that a single distribution be used where the data support it. A reasonable approach for the annual precipitation data would

Table 9.2. *Goodness of fit for the U.S. annual precipitation data.*

Distribution	Number acceptable
Generalized logistic	15
Generalized extreme-value	83
Lognormal	92
Pearson type III	92
Generalized Pareto	0

Note: The tabulated quantity is the number of homogeneous regions for which each of five distributions gave an acceptable fit to the region's annual precipitation totals.

be to use the lognormal distribution or the Pearson type III distribution for quantile estimation in the 92 regions where they both give an acceptable fit. In the other 16 homogeneous regions, for some of which none of the four distributions gave an acceptable fit, we prefer to use a distribution that will give good quantile estimates for as wide as possible a range of the true regional frequency distribution. The kappa and Wakeby distributions are robust to misspecification of the form of the frequency distribution in regional frequency analysis, as illustrated by the simulation results in Subsection 7.5.9, and either would be a good choice.

For the three heterogeneous regions, no single distribution is likely to give a good fit to each site's data, and at-site frequency analysis may be preferred to regional frequency analysis. However, regional frequency analysis may still be advantageous for estimation of extreme quantiles, as evidenced by the simulation results in Subsection 7.5.7 (particularly Figures 7.17 and 7.18). In these circumstances it is again advisable to use a robust distribution such as the kappa or Wakeby.

The National Drought Atlas used this approach – the analysis is described by Guttman et al. (1993). Counts were made by duration, region, and starting month of the number of times a distribution was acceptable. The Pearson type III was found to be acceptable most often for precipitation totals over all durations. It was acceptable for about 80% of all combinations of region, duration, and starting month. The lognormal and generalized extreme-value distributions were acceptable almost as often as the Pearson type III for durations longer than 6 months. Based on the counts of acceptable fits, the Pearson type III distribution was chosen for use in homogeneous regions for which it gave an acceptable fit, and the Wakeby distribution was used in the other homogeneous regions and in all heterogeneous regions.

9.1.6 Zero data values

In arid areas, monthly precipitation totals are often exactly zero and distributions fitted to them may take negative values unless the distribution is explicitly constrained to have a lower bound of zero. Negative quantile values violate the physical lower bound of zero for precipitation amounts. In the analysis of precipitation data for the National Drought Atlas, some of the estimated quantiles for regions in arid areas were negative for durations of three months and less in regions in arid areas. Although not a concern in our analysis of annual precipitation totals, we mention the problem here because it may occur with many kinds of data and the solution used in the National Drought Atlas is of general applicability.

The problem was solved by fitting to the regional data a mixed distribution with cumulative distribution function

$$F(x) = \begin{cases} 0, & x < 0, \\ p + (1-p)G(x), & x \geq 0. \end{cases} \tag{9.1}$$

Here $F(.)$ is the cumulative distribution function of precipitation amounts, p is the probability that the precipitation amount is zero, and $G(.)$ is the cumulative distribution function of the distribution of nonzero precipitation amounts. The parameter p was estimated by the proportion of zero values in the data for the region, and the parameters of the distribution G were estimated from the regional average L-moments of the nonzero data values.

As stated previously, the distribution G was initially chosen to be Pearson type III in homogeneous regions for which the Pearson type III distribution was accepted by the goodness of fit criterion and Wakeby otherwise. However, G was constrained to have a lower bound of zero when this was necessary to obtain nonnegative quantiles for all the probabilities of interest, the lowest of these being 0.02. When constrained estimation was necessary, the Wakeby with fixed lower bound $\xi = 0$ was fitted. A Pearson type III distribution with zero lower bound was not used because it has only two free parameters and rarely gave a good fit to the data. The lognormal and generalized extreme-value were not reconsidered as suitable distributions because the Wakeby can mimic the shapes of these distributions. The algorithm that was used to choose a distribution is shown in Figure 9.6.

9.1.7 Quantile estimates

Estimation of the regional frequency distribution and its quantiles for each region followed the regional L-moment algorithm described in Section 6.2, the fitted distribution being Pearson type III or Wakeby as explained above.

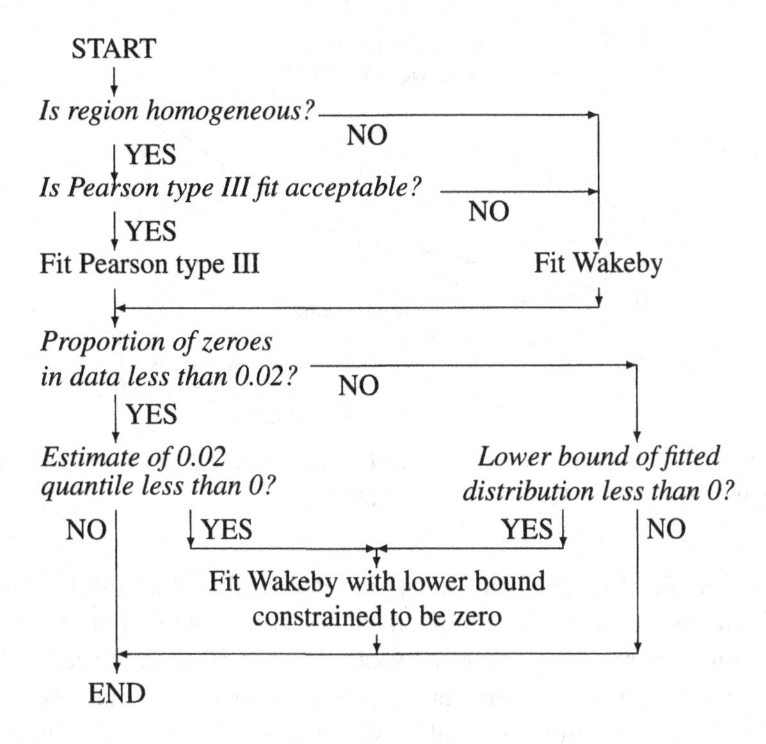

Fig. 9.6. Algorithm for choosing a distribution when the data contain exact zero values.

As an example of the results, Figure 9.7 shows the regional frequency distributions for two extreme regions. The regions' locations are indicated by the circles on Figure 9.8. One region, in central California, has an arid climate; its nine sites have mean annual precipitation between 6 in and 30 in. The regional frequency distribution is a Pearson type III distribution with $\tau = 0.198$ and $\tau_3 = 0.145$. The other region, in Washington state in the northwest, has a mild climate with relatively little variation in annual precipitation. Its eleven sites have mean annual precipitation between 38 in and 80 in. The regional frequency distribution is a Pearson type III distribution with $\tau = 0.095$ and $\tau_3 = 0.007$. Figure 9.7 is plotted as though on Normal probability paper. Return periods are indicated for events in both tails of the distribution. A Normal distribution would plot as a straight line on the graph. The Pearson type III distribution for the northwest Washington region has very low skewness and closely resembles a Normal distribution, and its plot is very close to being a straight line.

As a further illustration, Figure 9.8 maps for each region the regional growth factor $q(0.02)$, the magnitude of the extreme low event with a return period of 50 years. The numbers are expressed as percentages of the mean annual precipitation.

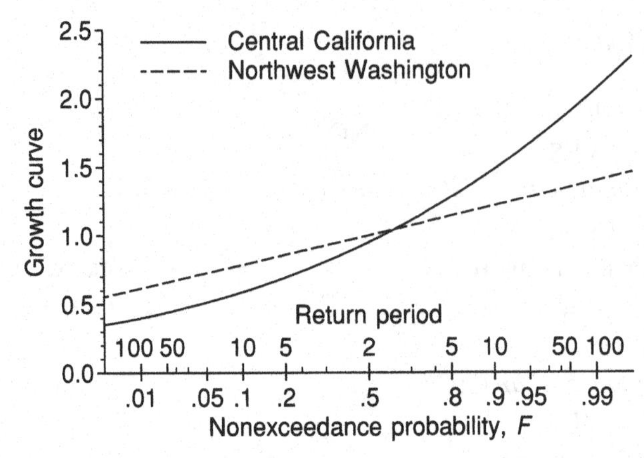

Fig. 9.7. Regional growth curves for two of the U.S. annual precipitation regions. Locations of the regions are indicated by the circles in Figure 9.8.

Thus, for example, the circled numeral 44 in central California indicates that, for a site in this region, an annual precipitation total as small as 44% of the mean annual precipitation is estimated to occur on average once in 50 years. Except for some mountainous areas in the western states, the percentages vary smoothly in space. This would be expected from physical considerations and suggests that the regional *L*-moment procedures do indeed capture the effects of the physical processes that operate in each region.

9.1.8 Accuracy of estimates

Quantile values were assessed by their bias and RMSE. These quantities cannot be calculated analytically because the regional *L*-moment quantile estimation procedure is too complicated. Instead, a Monte Carlo simulation procedure was used, as described in Section 6.4. Simulated data were generated for a region with the same number of sites and the same record lengths as the actual region and were drawn from the distribution that was fitted to the actual regional data. Because the correlation between sites is substantial, typically between 0.4 and 0.8, the simulations used correlated data generated by the algorithm described in Table 6.1. Quantile estimates were calculated for the sites in this simulated region. The simulation was repeated 500 times. The 500 sets of errors in the simulated quantile estimates were accumulated and averaged to yield approximations to the bias and RMSE of the quantile estimates calculated from the actual data.

As an example, Figure 9.9 shows RMSEs of the regional quantile $q(0.02)$ of the annual precipitation totals; these values may be regarded as the standard errors of the estimates mapped in Figure 9.8. The RMSEs are certainly small enough to enable the quantile estimates to be used with confidence.

9.1.9 Summary

Computation of quantile values for the National Drought Atlas was the first known large-scale application of regional frequency analysis using *L*-moments. We judge it to be a success for the following reasons: homogeneous and physically reasonable regions were defined for almost all of the study area; the observed patterns of estimated quantiles obtained from regional frequency analysis generally show little variation between adjacent regions, and are in good agreement with prior climatological expectations; and the accuracy of the final quantile estimates, as measured by the estimated RMSEs, is satisfactorily low.

The analysis also illustrates some of the typical difficulties of regional frequency analysis. The most important of these is obtaining homogeneous regions, which required a time-consuming process of detailed inspection of the data and consideration of the factors that define a precipitation climate.

From a climatological point of view, the number of regions resulting from this study may be considered excessive. Fovell and Fovell (1993), for example, defined 25 climate zones from cluster analysis of monthly average temperature and precipitation data. A larger number of regions is appropriate for regional frequency analysis, because geographically compact regions are required to capture local variations in the distribution of precipitation totals within zones that have generally similar climates.

9.2 Annual maximum streamflow in central Appalachia

9.2.1 Data

Smith (1992) analysed data on annual maximum streamflow at 104 gaging stations in the central Appalachia region of the United States. Smith was interested in relating the first two moments of the at-site frequency distributions to the drainage area of the basins. We use the same set of sites to illustrate regional frequency analysis when measurements are available for several site characteristics including one, in this case drainage basin area, that is known to have a particularly strong influence on the at-site frequency distribution. Data were obtained from "Hydrodata" CD-ROMs (Hydrosphere, 1993), which reproduce data from the U.S. Geological Survey's WATSTORE data files.

The study area includes parts of the Piedmont and Valley and Ridge physiographic provinces of Virginia and Maryland. Locations of the sites are shown in Figure 9.10.

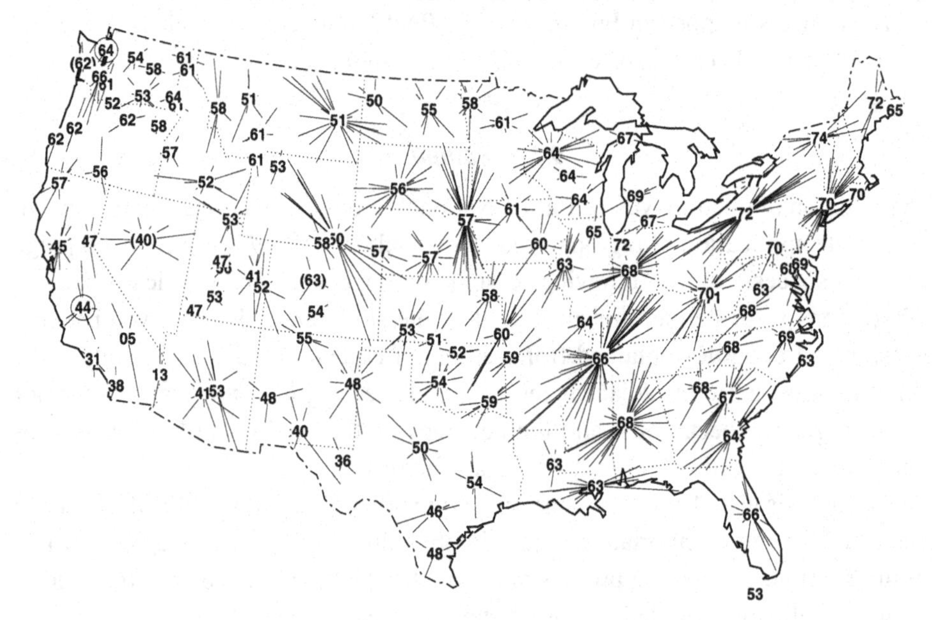

Fig. 9.8. Estimated regional growth factor at return period 50 years in the lower tail of the distribution, $q(0.02)$, expressed as a percentage of the mean. Figures are plotted at the center of each region. Figures in parentheses indicate that the region is heterogeneous. Lines radiating from the center end at the locations of stations in the region. Circles identify the regions whose regional growth curves are given in Figure 9.7.

Drainage areas of the basins cover a very wide range, from 0.3 mi^2 to 10,000 mi^2. Elevations of the gaging sites range from 10 ft to 2,000 ft above sea level. In addition to these variables, the data set contains the gage location (latitude and longitude) and the magnitude and date of the maximum instantaneous streamflow in each "water year," running from Oct. 1 through Sep. 30, for some or all of the years 1895–1991. Some variables that might be expected to influence the frequency distribution, such as mean annual precipitation over the basin, the underlying geology of the basin, and the extent to which the basin is forested or urbanized, are not available.

Two gage elevations are missing from the data set, and were estimated from maps. Some streamflow values, 26 in all, were missing and were ignored. There are 24 data points marked as being "historic values." All but one of these are separated from the main sequence of the gaged record by one or more missing years. We suspect that these values are present in the data only because their magnitude is exceptionally large; therefore their inclusion in the data would prevent the data from being regarded as a random sample. Accordingly, they were excluded from the data set. The other historic value occurred at the start of a continuous sequence of gaged record and was retained. Site 01626000 has an observation in 1943 but

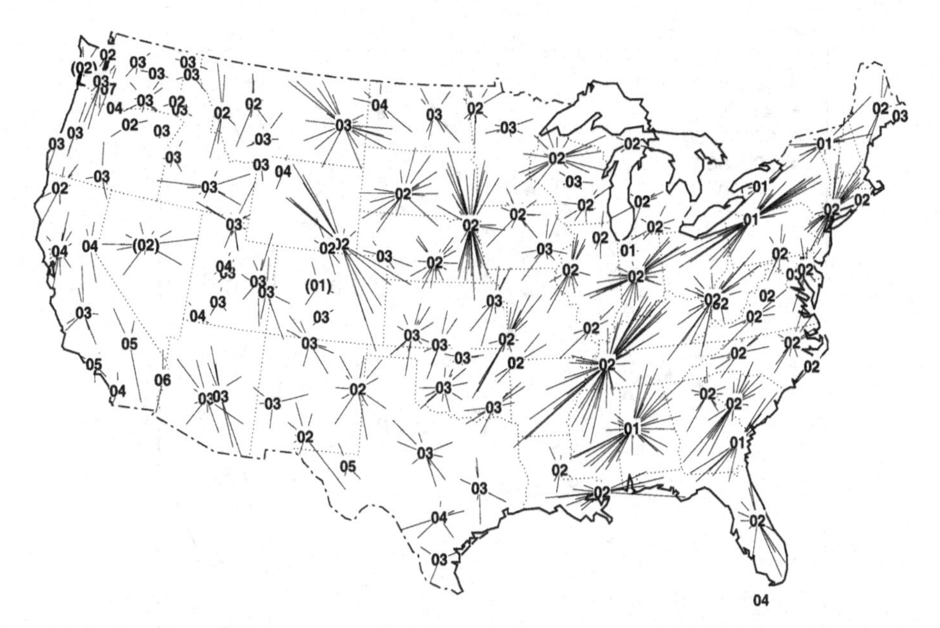

Fig. 9.9. Estimated RMSE of the regional growth factor at return period 50 years in the lower tail of the distribution, $q(0.02)$, expressed as a percentage of the mean. Figures are plotted at the center of each region. Figures in parentheses indicate that the region is heterogeneous. Lines radiating from the center end at the locations of stations in the region.

no others until a continuous period of gaging started in 1953. The value for 1943 is unusually high and appears to be a historic value that was not marked as such. We deleted it from the data set, again because we suspect that it is present only because its magnitude is exceptionally large.

With these modifications, the data set contains a total of 4,456 streamflow observations, with sample sizes at the 104 sites varying from 19 to 97. The at-site sample L-CV ranges from 0.22 to 0.71, and the at-site sample L-skewness ranges from 0.14 to 0.74, except that one site has the exceptionally low value −0.18. The average value for L-CV is 0.42 and for L-skewness is 0.44. These are very high values, compared for example with those of the annual precipitation total data in Figure 9.1, and indicate that the frequency distributions are highly skew.

9.2.2 Initial screening of data

The L-CV and L-skewness of the data are shown in Figure 9.11. Treating the entire set of 104 sites as a single region, the discordancy statistic D_i of Eq. (3.3) was calculated for each site.

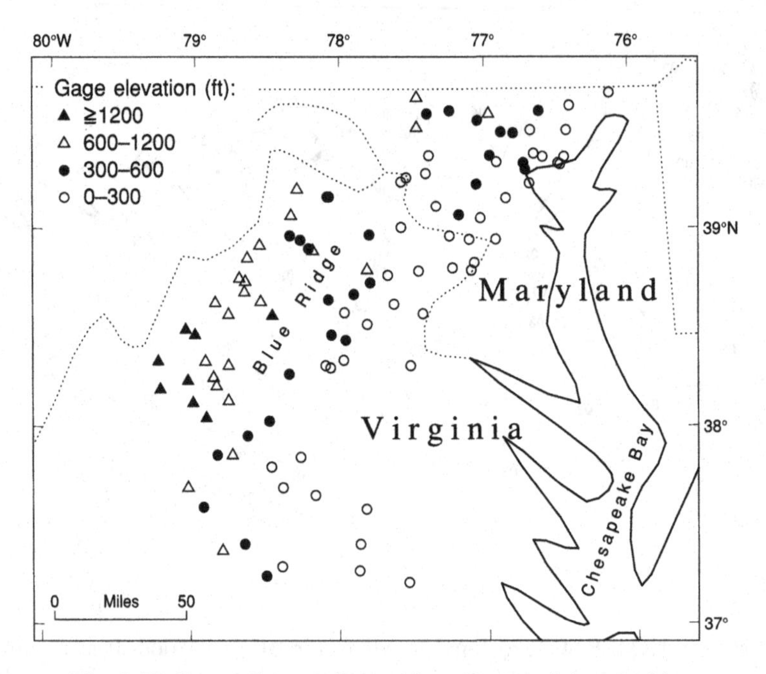

Fig. 9.10. Streamflow gaging stations in central Appalachia.

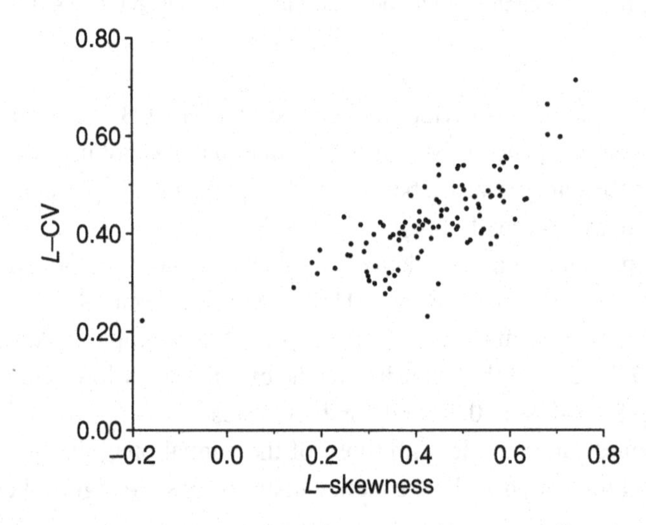

Fig. 9.11. *L*-moment ratios for the Appalachian streamflow data.

One site stands out as being particularly discordant, with $D_i = 13.5$. This is the site with negative *L*-skewness. The data for this site, site 01624800, are shown in Figure 9.12, together with the data for the site's nearest neighbor, site 01626000. There is good qualitative agreement between the sites: at both, the data values

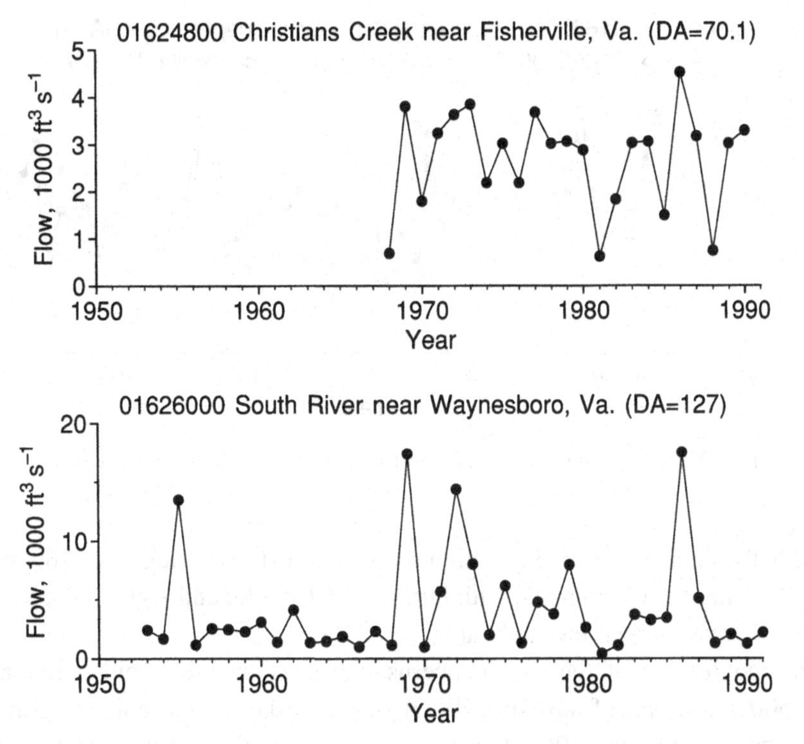

Fig. 9.12. Annual maximum flows for site 01624800 and its nearest neighbor. DA denotes drainage area, in square miles.

for 1968, 1981, and 1988 are among the lowest and those for 1969 and 1986 are high. To some extent the negative *L*-skewness at site 01624800 occurs because the period of gaging included three years that had particularly low maximum flows. However, there are also physical reasons that may explain the pattern of the data at this site. The geology of the basin consists of fractured quartzite sandstone overlying dolomite, resulting in a Karstlike hydrology; the basin is very porous and flood response to even the most extreme storm events is slow. In consequence, the frequency distribution of annual maximum flows at the gaging site should be less skew than at sites with impervious geology and a fast response time to extreme storms. It is not clear whether this site is geologically so atypical of the rest of the data that it should be excluded from the analysis. We therefore decided to retain the site in the data set. As noted below, this makes little difference to the results of the regional frequency analysis.

Other checks on the data were inconclusive. There are no obvious trends or changes in level apparent in thc data. Some sites have unusually high sample *L*-CV and *L*-skewness, but the data appear to be correct. At the four sites with *L*-CV greater

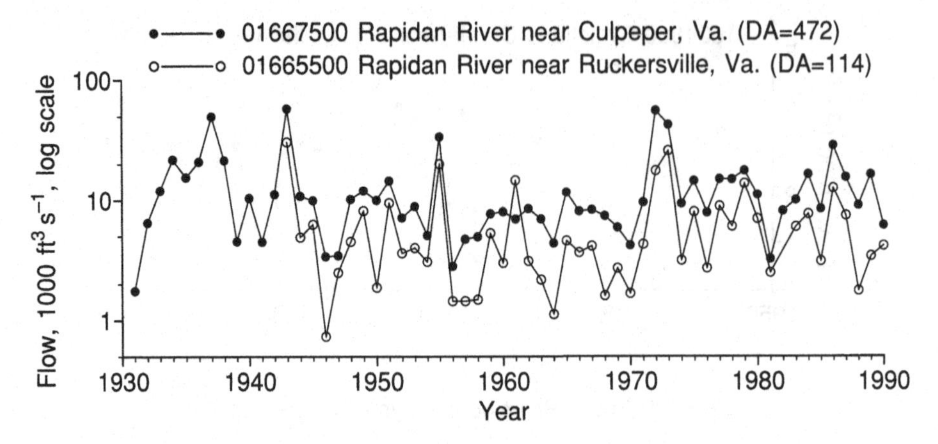

Fig. 9.13. Annual maximum flows for two sites on the Rapidan River.

than 0.6, the data contain a single high outlier occurring on Aug. 20, 1969 or June 21, 1972. These dates coincide with hurricanes (Camille and Agnes, respectively) that passed over or near the study area.

One odd feature of the data concerns two sites on the Rapidan River. Site 01665500 is upstream of site 01667500. A plot of the data for these sites, Figure 9.13, shows that in one year, 1961, the maximum flow at the upstream site exceeded that at the downstream site. Although physically possible, such an occurrence is sufficiently odd that it would be worth checking the validity of the data for that year at these sites. Lacking easy access to sufficiently detailed data for such a check, we retained both sites in the data set.

9.2.3 Formation of regions

Treating the entire set of 104 sites as a single region, the heterogeneity statistic (4.5) was evaluated as $H = 2.08$. The entire set is therefore not far from being homogeneous, or at worst "possibly heterogeneous". Nonetheless, we emphatically reject the possibility of performing regional frequency analysis with the entire set of sites being treated as a single region. The main reason is that the theory and practice of hydrology imply that the frequency distribution is likely to depend on the drainage area of the basin. Regional frequency analysis should therefore be applied only to regions whose basins cover a fairly small range of drainage area. A further point is that in regional frequency analysis there is little to be gained by using regions containing more than about 20 sites. A reasonable starting point for regional frequency analysis would therefore be a subdivision of the set of sites, according to their drainage areas, into groups of not much more than 20.

Table 9.3. *Summary of initial regions for Appalachian streamflow data.*

Region	Range of area	Number of sites	Regional average t	Regional average t_3	H	Acceptable fit
A	< 10	26	0.423	0.402	2.88	GLO GEV
B	10–60	28	0.450	0.472	1.70	GLO GEV
C	60–200	27	0.430	0.459	0.03	none
D	> 200	23	0.387	0.419	0.29	GLO GEV

Note: Regions are defined in terms of drainage area alone. Units of drainage area are square miles. H is the heterogeneity statistic defined in Eq. (4.5). "Acceptable fit" indicates which of five distributions gave a value $|Z| \leq 1.64$ for the Z statistic defined in Eq. (5.6). The five distributions were generalized logistic (GLO), generalized extreme-value (GEV), lognormal, Pearson type III, and generalized Pareto.

The set of sites was accordingly divided into four groups of approximately equal size according to the sites' drainage areas. We term these groups Regions A–D. The discordancy, heterogeneity, and goodness of fit measures described in Chapters 3–5 were applied to each region's data. A summary of the results is contained in Table 9.3. The heterogeneity measure H indicates that homogeneous regions were achieved for sites with drainage area greater than 60 mi^2 but that the regions containing the smaller basins were heterogeneous. Other attempts to define homogeneous regions based on drainage area alone gave similar results.

It appears that, at least for the smaller basins, the frequency distribution of annual maximum streamflow is determined by more factors than drainage area alone. Accordingly, more information was introduced into the procedure for forming regions. The four available site characteristics, drainage basin area, gage elevation, gage latitude, and gage longitude, were used in a cluster analysis procedure. Nonlinear transformations were applied to two of the variables: a logarithmic transformation to drainage basin area and a square root transform to gage elevation. These transformations give a more symmetric distribution of the values of the site characteristics at the 104 sites, reducing the likelihood that a few sites will have site characteristics so far from the other sites that they will always be assigned to a cluster by themselves, and, in our judgement, give a better correspondence between differences in site characteristics and the degree of hydrologic dissimilarity between different basins. All four variables were then standardized by dividing by the standard deviation of their values at the 104 sites. Finally, the drainage basin area variable was multiplied by 3 to give it an importance in the clustering procedure equal to that of the other variables together. The transformed variables used in the cluster analysis are listed in Table 9.4. The transformation and weighting of the

Table 9.4. *Transformation of site characteristics.*

Site characteristic, X	Cluster variable, Y
Drainage basin area (mi^2)	$Y = \log(X) \times 3/\text{s.d.}(\log X)$
Gage elevation (ft)	$Y = \sqrt{X}/\text{s.d.}(\sqrt{X})$
Gage latitude (deg)	$Y = X/\text{s.d.}(X)$
Gage longitude (deg)	$Y = X/\text{s.d.}(X)$

Note: Here s.d.(X) denotes the standard deviation of the 104 values of the site characteristic X.

variables involves subjective decisions whose justification is the physical plausibility of the regions that are ultimately obtained from the clustering procedure.

Cluster analysis was performed using Ward's method. This is an "agglomerative hierarchical" clustering procedure. Initially each site is a cluster by itself, and clusters are then merged one by one until all sites belong to a single cluster. The assignment of sites to clusters can be determined for any number of clusters, and there is no formal measure of an "optimal" number of clusters. Choice of a suitable number of clusters is therefore subjective. The number of sites in a cluster should be large, to obtain the maximum benefit of regionalization, but the range of drainage areas of the sites in a cluster should not be too large, for otherwise it would not be reasonable to expect the clusters to be homogeneous. For the Appalachian streamflow sites we judged that seven clusters would be an appropriate number. The clusters obtained by Ward's method were adjusted using the K-means algorithm of Hartigan and Wong (1979), which yielded clusters that were a little more compact in the space of cluster variables.

Figure 9.14 shows the final clusters on a graph whose axes are the transformed drainage basin area and gage elevation variables. Clusters are numbered 1–7 in increasing order of average drainage basin area for the sites in the cluster. Region 1 contains three sites with very small drainage area. Regions 2, 4, 5, and 7 span the range of drainage area from 2 mi^2 to 10,000 mi^2 and contain mostly low-elevation sites. Regions 3 and 6 contain mostly high-elevation sites. These distinctions are not exact, because the remaining variables, latitude and longitude, also affect the clustering. Their effect can be seen in Figure 9.15, which shows the geographical location of the sites in each cluster. Regions 2 and 5 lie predominantly to the northeast and Regions 3 and 6 lie to the west of the study area.

This set of clusters is not completely satisfactory, because we lack data on some useful site characteristics, such as urbanization. In an urban area, much of the land is covered by an impervious layer of concrete or asphalt and stream channels are often artificially constructed and straighter than natural channels. An urban basin's flood

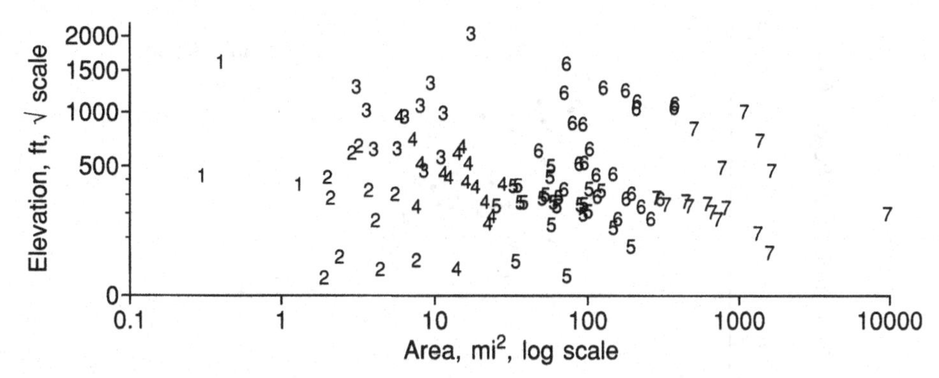

Fig. 9.14. Drainage basin area and gage elevation for clusters of streamflow gaging stations in central Appalachia.

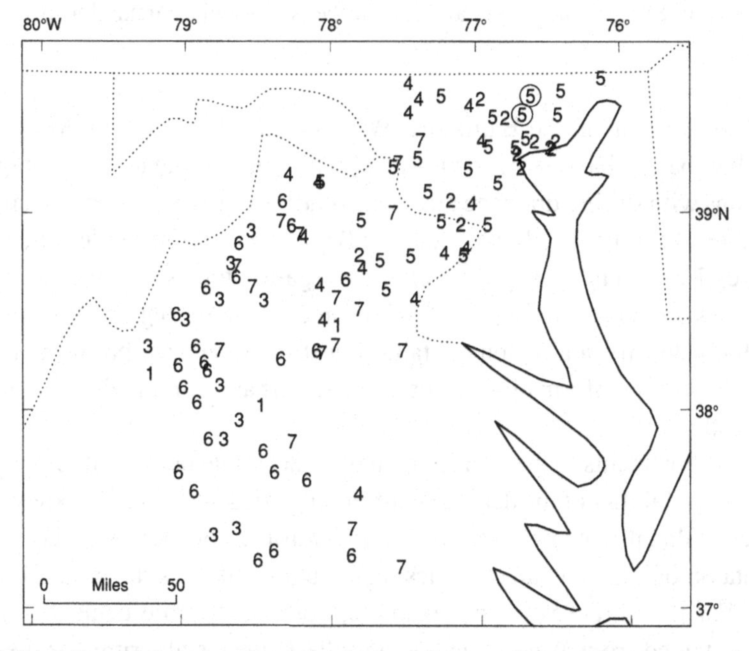

Fig. 9.15. Location of the sites in each cluster of streamflow gaging stations in central Appalachia. The two circled sites are discussed further in Subsection 9.2.5.

response to storms, and the distribution of its maximum streamflows, is therefore likely to be different from that of an otherwise comparable basin in a rural area less affected by man's control of the environment. One difference between urban and rural basins is in their specific mean annual flood, defined to be the mean annual maximum streamflow divided by the drainage area of the basin. Figure 9.16 is a plot of specific mean annual flood for the Appalachian basins, identifying the points that

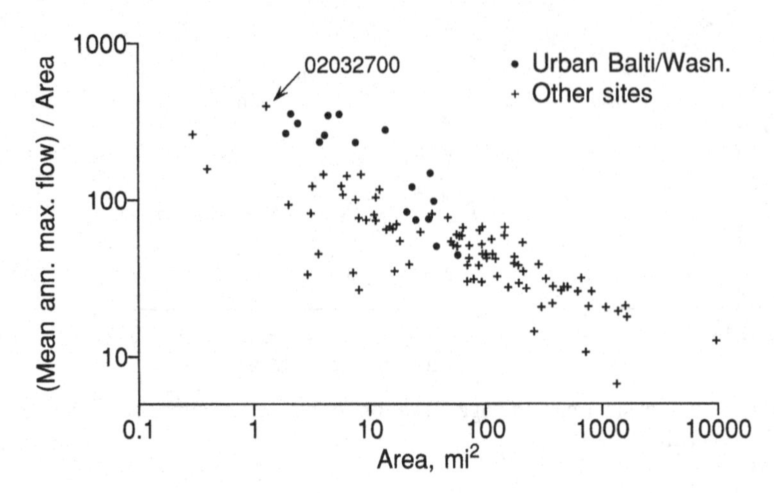

Fig. 9.16. Specific mean annual flood for the Appalachian streamflow basins.

correspond to basins in the Baltimore–Washington urban corridor. Most of these small urban basins have large values of specific mean annual flood compared to other basins with similar drainage area. One other site also has a large specific mean annual flood. This site, 02032700, Schenks Branch at Charlottesville, Va., is a small basin (area 1.3 mi^2) in a fairly large city; we regard it too as an urban basin. These urban basins are mostly in Region 2, with some others in Regions 1 and 4. It may be worthwhile to form a region containing just the urbanized basins, but we lack sufficiently precise information on the extent of urbanization in different basins to be sure of this.

Further adjustments to these clusters may be desirable but would require further study of the reliability of the data from different gaging sites, detailed knowledge of local factors that might affect the climate or hydrology of each drainage basin, and more data on other site characteristics that could be used as the basis for forming regions. For the purposes of this example, however, we are content to take the clusters obtained from Ward's method and the K-means algorithm as the regions for use in frequency analysis.

The discordancy, heterogeneity, and goodness-of-fit measures described in Chapters 3–5 were applied to each region's data. Summary statistics for the regions are contained in Table 9.5. The heterogeneity measure H indicates that homogeneous regions were achieved in Regions 5–7, containing sites with drainage area greater than 30 mi^2. Regions containing the smaller basins were not so successful but should still be satisfactory for regional frequency analysis. Regions 2–4 are "possibly heterogeneous", with H values as high as 1.69, but should yield quantile estimates more accurate than those obtained from single-site frequency analysis.

Table 9.5. *Summary of final regions for Appalachian streamflow data.*

Region	Range of area	Number of sites	Regional average t	Regional average t_3	H	Acceptable fit
1	0.3–1.3	3	0.348	0.331	−0.40	GLO GEV LN3 GPA PE3
2	1.9–7.6	11	0.379	0.363	1.65	GLO GEV LN3
3	3.1–17	11	0.498	0.471	1.62	GLO GEV LN3
4	5.9–28	16	0.450	0.430	1.69	GLO GEV LN3 GPA
5	25–190	23	0.424	0.491	0.24	none
6	47–380	24	0.438	0.442	−0.03	GLO GEV
6′	47–380	23	0.442	0.455	−0.71	GLO GEV
7	290–9600	16	0.376	0.420	0.87	GLO GEV

Note: Regions 1–7 are those obtained by cluster analysis. Region 6′ is Region 6 with site 01624800 excluded. Units of drainage area are square miles. H is the heterogeneity statistic defined in Eq. (4.5). "Acceptable fit" indicates which of five distributions gave a value $|Z| \le 1.64$ for the Z statistic defined in Eq. (5.6). The five distributions were generalized logistic (GLO), generalized extreme-value (GEV), lognormal (LN3), Pearson type III (PE3), and generalized Pareto (GPA).

Regions 6 and 7 each contain one discordant site. Region 6 contains the site with negative L-skewness, site 01624800, for which $D_i = 6.9$. As discussed previously, it is not clear whether this site should be deleted from the data set. To do so makes little difference to the results: Region 6′, defined as consisting of all sites in Region 6 except 01624800, is also homogeneous and its regional average L-moment ratios are very similar to those of Region 6. However, quantile estimates at the discordant site itself may be unreliable, and at-site estimation may be preferable there. In Region 7, site 02041500 has low L-CV, $t = 0.23$, and $D_i = 3.5$. Examination of the data shows no exceptional features, so the site was retained in the data set.

9.2.4 Choice and estimation of the frequency distribution

Figure 9.17 is an L-moment ratio diagram showing the regional average L-skewness and L-kurtosis of the seven final regions. Consistently with the results in Table 9.5, most of the points in Figure 9.17 lie close to the generalized extreme-value and generalized logistic lines. The lognormal distribution, used by Smith (1992), is acceptable in Regions 1–4, but is rejected for the regions containing large-area basins. If the use of a single distribution for each region is desired, the generalized logistic or generalized extreme-value, or a robust distribution such as the kappa or Wakeby, would be a reasonable choice. Table 9.6 shows the quantiles of the regional frequency distributions, obtained by fitting generalized extreme-value distributions to each region's data using the regional L-moment algorithm described in Chapter 6.

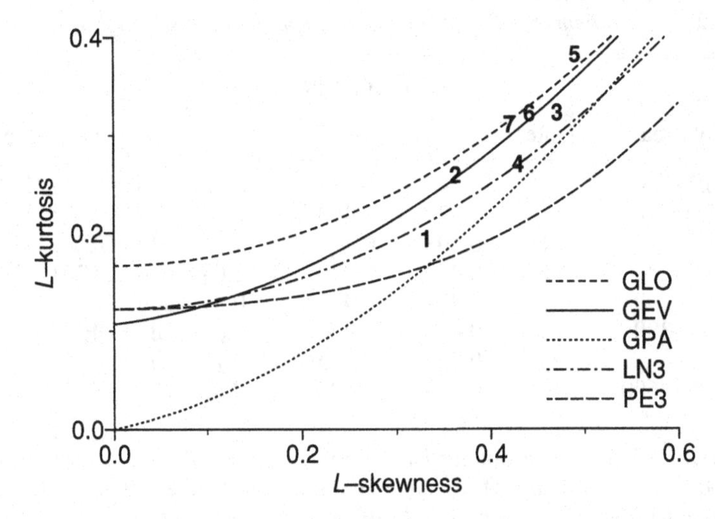

Fig. 9.17. Regional average *L*-moments for the Appalachian streamflow regions.

Table 9.6. *Estimated quantiles for Appalachian streamflow regions.*

	Nonexceedance probability									
Region	0.01	0.10	0.20	0.50	0.80	0.90	0.98	0.99	0.998	0.999
1	0.17	0.37	0.49	0.81	1.35	1.80	3.12	3.85	6.09	7.34
2	0.14	0.34	0.45	0.78	1.35	1.85	3.39	4.29	7.18	8.88
3	0.03	0.20	0.31	0.64	1.32	1.99	4.46	6.15	12.60	17.04
4	0.07	0.25	0.37	0.70	1.34	1.95	4.03	5.38	10.18	13.30
5	0.20	0.33	0.42	0.69	1.25	1.82	3.98	5.50	11.47	15.69
6	0.11	0.28	0.39	0.70	1.32	1.91	3.98	5.34	10.30	13.57
7	0.22	0.37	0.47	0.75	1.30	1.80	3.52	4.61	8.45	10.91

Note: Fitted distribution, generalized extreme-value.

The quantiles for different regions are quite similar in the main body of the distribution; only at the $F = 0.98$ quantile and beyond do marked differences become apparent. Regions 1 and 2 have relatively low upper-tail quantiles. Otherwise the upper-tail quantiles generally increase as the drainage area of the basin decreases, though Region 4 is an exception to this pattern.

9.2.5 Accuracy of estimation

Error bounds for the estimated regional growth curves were obtained by simulation, using the method described in Section 6.4. The simulations assumed that the regions were homogeneous with generalized extreme-value frequency distributions. There

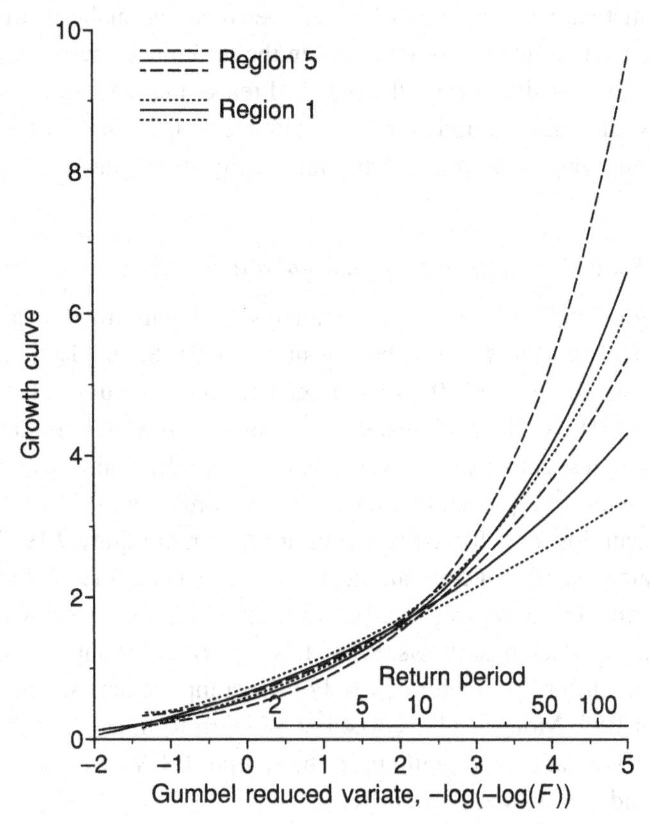

Fig. 9.18. Estimated regional growth curves, with their 90% error bounds, for Appalachian streamflow regions 1 and 5.

is no evidence of intersite dependence in Region 1, but intersite dependence is present in the data for the other regions, the average correlation between sites varying from 0.30 in Region 2 to 0.67 in Region 5. The simulations for Regions 2–7 therefore included intersite dependence, using the algorithm described in Table 6.1. Results for Regions 1 and 5, which have the lowest and highest *L*-skewness of the six regions, are shown in Figure 9.18. Error bounds are given only for exceedance probabilities of 0.02 or greater, because for extreme low quantiles the simulated estimates were sometimes negative and led to unreliable error bounds, as noted in Section 6.4.

Although the regional growth curves for Regions 1 and 5 are fairly well separated, for other regions there is considerable overlap between the error bounds of the regional growth curves. This suggests that on statistical grounds there is little justification for treating the regions as distinct and that they might as well be merged to form larger regions. This argument should be treated with caution, because the absence of statistically significant differences between regional growth curves may merely reflect an insufficiency of data. When the differences between the regional growth curves are consistent with the physical reasoning that led to the definition of the regions, there are grounds for belief that these differences are scientifically significant even though they may not be statistically significant.

9.2.6 A comparison of regional and at-site estimation

The reasoning behind regional frequency analysis is exemplified by a comparison of frequency analyses for two neighboring sites: site 01582000, Little Falls at Blue Mount, Md., and site 01583500, Western Run at Western Run, Md. The sites are circled on Figure 9.15. The gaging sites are 7 mi apart and have similar elevations, 305 ft and 262 ft, respectively; the basins also have similar drainage areas, 52.9 mi^2 and 59.8 mi^2, respectively. Both sites have gaged records from 1945 to 1991. Annual maximum streamflow data for the two sites are shown in Figure 9.19. The data for the two sites are generally of comparable magnitude except in 1972. At both sites in that year the annual peak was recorded on June 22, 1972, coincident with Hurricane Agnes, but the recorded peak flows, 8,280 $ft^3 s^{-1}$ at Blue Mount and 38,000 $ft^3 s^{-1}$ at Western Run, differ by a factor of 4.6. For the complete data samples, the *L*-CV is $t = 0.30$ at Blue Mount and $t = 0.49$ at Western Run. Most of the difference is due to the 1972 data value: without it, the sample *L*-CVs would be $t = 0.28$ at Blue Mount and $t = 0.36$ at Western Run.

Table 9.7 gives the results of frequency analyses for the two sites. In the at-site analyses a generalized extreme-value distribution was fitted at each site using the method of *L*-moments. The estimated quantiles in the upper tail of the frequency distribution differ by factors of 1.4, 2.5, and 4.9 at nonexceedance probabilities $F = 0.9$, $F = 0.99$, and $F = 0.999$, corresponding to return periods of 10, 100, and 1,000 years. The Bulletin 17 analyses followed the procedure described in Section 8.3, using a generalized skew coefficient $g_{map} = 0.6$. The computation was performed using the HEC-FFA program (U.S. Army Corps of Engineers, 1992). The results are very similar to the at-site analysis based on *L*-moments, except that the estimates for Western Run at $F = 0.998$ and $F = 0.999$ are somewhat lower. The regional analysis used the regional growth curve of Region 5 in Table 9.6, to which both sites belong, and estimated the index flood at each site by the sample mean. Because the ratio of the sample means at the two sites is 1.29, the quantile estimates differ by this factor at all return periods.

These quantile estimates raise a critical question for frequency estimation of extreme events. Is the greater flood response of the Western Run basin a systemic property that can be expected to recur whenever future large storms pass over the study area, or are the different responses of the two basins a result of unpredictable

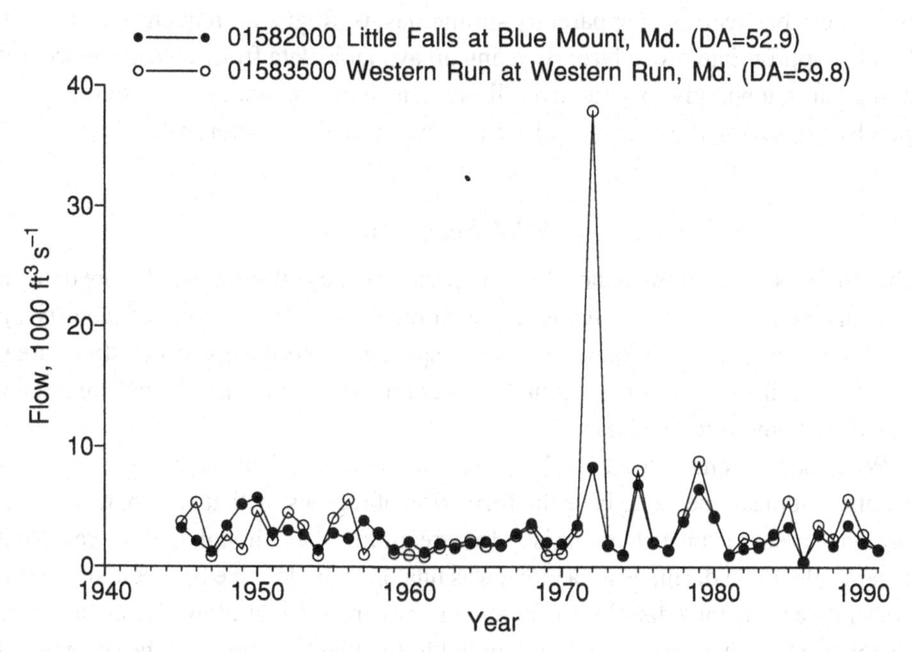

Fig. 9.19. Annual maximum streamflow data for two sites in central Appalachian streamflow Region 5.

Table 9.7. *Estimated quantiles of annual maximum streamflow, in cubic feet per second, for Blue Mount and Western Run.*

		Nonexceedance probability				
Method	Site	0.9	0.98	0.99	0.998	0.999
At-site	Blue Mount	4660	7570	9110	13600	16000
	Western Run	6430	15800	22990	54100	78000
Bulletin 17	Blue Mount	4680	7680	9270	13800	16300
	Western Run	6740	15800	21900	45100	60600
Regional	Blue Mount	4960	10900	15000	31300	42800
	Western Run	6390	14000	19300	40300	55100

local variations in precipitation intensity and duration, with future storms being as likely to cause greater runoff in one basin as in the other? At-site analysis is appropriate if the former explanation is correct, though even so it will be accurate only if the frequency of extreme events in the observed record matches the long-term rate of occurrence of such events. We believe, however, that differences so large as that illustrated in Figure 9.19 can rarely be ascribed with confidence to physical differences between such apparently similar basins. Regional frequency analysis is then the appropriate procedure. By using all available data from a set of physically similar sites, it will give a better overall estimate of the frequency of extreme events, thereby improving the accuracy of frequency estimation at every site.

9.2.7 Summary

This analysis of streamflow data does not pretend to be authoritative; if more data on site characteristics were available, there would be a better prospect of identifying local variations in the climate or flood response of the basins in the study area. However, it illustrates some features that commonly occur in regional frequency analysis of environmental data.

When one site characteristic is known to have a strong effect on the frequency distribution, it makes sense to base the formation of regions on that site characteristic. Drainage area is such a characteristic for streamflow data, though in this example it became clear that drainage area alone was unable to explain the between-site variations in the frequency distribution of annual maximum streamflow. Cluster analysis closely tied to physical reasoning then led to the identification of other sources of variation between at-site frequency distributions. This approach should be widely applicable, though the patterns of dependence of the frequency distribution on site characteristics may be quite different in other parts of the world.

The main conclusion is that successful regional frequency analysis involves interaction between physical reasoning and statistical assessment of the regions and their estimated frequency distributions. Close study of the physical reasons for variation between sites is essential to the successful use of regional frequency analysis, because it enables reasonable regions to be obtained even when small samples or irregularities in the data make it difficult to define convincing regions by statistical methods alone.

Appendix

L-moments for some specific distributions

For each of the distributions listed here we give the form of the probability density function $f(x)$, the cumulative distribution function $F(x)$, and the quantile function $x(F)$, expressions for the L-moments in terms of the parameters and for the parameters in terms of the L-moments. The expressions for the parameters are used to calculate estimates of the parameters by the method of L-moments.

A.1 Uniform distribution

Definition

Parameters (2): α (lower endpoint of the distribution), β (upper endpoint).

Range of x: $\alpha \le x \le \beta$.

$$f(x) = 1/(\beta - \alpha) \tag{A.1}$$

$$F(x) = (x - \alpha)/(\beta - \alpha) \tag{A.2}$$

$$x(F) = \alpha + (\beta - \alpha)F \tag{A.3}$$

L-moments

$$\lambda_1 = \tfrac{1}{2}(\alpha + \beta) \tag{A.4}$$

$$\lambda_2 = \tfrac{1}{6}(\beta - \alpha) \tag{A.5}$$

$$\tau_3 = 0 \tag{A.6}$$

$$\tau_4 = 0 \tag{A.7}$$

Parameters

Parameter estimates for the uniform distribution are of little interest in regional frequency analysis and are therefore omitted.

A.2 Exponential distribution

Definition

Parameters (2): ξ (lower endpoint of the distribution), α (scale).
Range of x: $\xi \le x < \infty$.

$$f(x) = \alpha^{-1} \exp\{-(x-\xi)/\alpha\} \tag{A.8}$$

$$F(x) = 1 - \exp\{-(x-\xi)/\alpha\} \tag{A.9}$$

$$x(F) = \xi - \alpha \log(1-F) \tag{A.10}$$

L-moments

$$\lambda_1 = \xi + \alpha \tag{A.11}$$

$$\lambda_2 = \tfrac{1}{2}\alpha \tag{A.12}$$

$$\tau_3 = \tfrac{1}{3} \tag{A.13}$$

$$\tau_4 = \tfrac{1}{6} \tag{A.14}$$

Parameters

If ξ is known, α is given by $\alpha = \lambda_1 - \xi$ and the L-moment, moment, and maximum-likelihood estimators are identical. If ξ is unknown, the parameters are given by

$$\alpha = 2\lambda_2, \qquad \xi = \lambda_1 - \alpha. \tag{A.15}$$

For estimation based on a single sample these estimates are inefficient, but in regional frequency analysis they can give reasonable estimates of upper-tail quantiles.

A.3 Gumbel (extreme-value type I) distribution

Definition

Parameters (2): ξ (location), α (scale).
Range of x: $-\infty < x < \infty$.

$$f(x) = \alpha^{-1} \exp\{-(x-\xi)/\alpha\} \exp[-\exp\{-(x-\xi)/\alpha\}] \tag{A.16}$$

$$F(x) = \exp[-\exp\{-(x-\xi)/\alpha\}] \tag{A.17}$$

$$x(F) = \xi - \alpha \log(-\log F) \tag{A.18}$$

L-moments

$$\lambda_1 = \xi + \alpha\gamma \tag{A.19}$$

$$\lambda_2 = \alpha \log 2 \tag{A.20}$$

$$\tau_3 = 0.1699 = \log(9/8)/\log 2 \tag{A.21}$$

$$\tau_4 = 0.1504 = (16\log 2 - 10\log 3)/\log 2 \tag{A.22}$$

Here γ is Euler's constant, 0.5772... .

Parameters

$$\alpha = \lambda_2/\log 2, \qquad \xi = \lambda_1 - \gamma\alpha \tag{A.23}$$

A.4 Normal distribution

Definition

Parameters (2): μ (location), σ (scale).

Range of x: $-\infty < x < \infty$.

$$f(x) = \sigma^{-1}\phi\left(\frac{x-\mu}{\sigma}\right) \tag{A.24}$$

$$F(x) = \Phi\left(\frac{x-\mu}{\sigma}\right) \tag{A.25}$$

$x(F)$ has no explicit analytical form

Here

$$\phi(x) = (2\pi)^{-1/2}\exp(-\tfrac{1}{2}x^2), \qquad \Phi(x) = \int_{-\infty}^{x} \phi(t)\,dt. \tag{A.26}$$

L-moments

$$\lambda_1 = \mu \tag{A.27}$$

$$\lambda_2 = 0.5642\sigma = \pi^{-1/2}\sigma \tag{A.28}$$

$$\tau_3 = 0 \tag{A.29}$$

$$\tau_4 = 0.1226 = 30\pi^{-1}\arctan\sqrt{2} - 9 \tag{A.30}$$

Parameters

$$\mu = \lambda_1, \qquad \sigma = \pi^{1/2}\lambda_2 \tag{A.31}$$

A.5 Generalized Pareto distribution

Definition

Parameters (3): ξ (location), α (scale), k (shape).

Range of x: $\xi \le x \le \xi + \alpha/k$ if $k > 0$; $\xi \le x < \infty$ if $k \le 0$.

$$f(x) = \alpha^{-1} e^{-(1-k)y}, \quad y = \begin{cases} -k^{-1}\log\{1 - k(x-\xi)/\alpha\}, & k \ne 0 \\ (x-\xi)/\alpha, & k = 0 \end{cases} \tag{A.32}$$

$$F(x) = 1 - e^{-y} \tag{A.33}$$

$$x(F) = \begin{cases} \xi + \alpha\{1 - (1-F)^k\}/k, & k \ne 0 \\ \xi - \alpha\log(1-F), & k = 0 \end{cases} \tag{A.34}$$

Special cases: $k = 0$ is the exponential distribution; $k = 1$ is the uniform distribution on the interval $\xi \le x \le \xi + \alpha$.

L-moments

L-moments are defined for $k > -1$.

$$\lambda_1 = \xi + \alpha/(1+k) \tag{A.35}$$

$$\lambda_2 = \alpha/\{(1+k)(2+k)\} \tag{A.36}$$

$$\tau_3 = (1-k)/(3+k) \tag{A.37}$$

$$\tau_4 = (1-k)(2-k)/\{(3+k)(4+k)\} \tag{A.38}$$

The relation between τ_3 and τ_4 is given by

$$\tau_4 = \frac{\tau_3(1+5\tau_3)}{5+\tau_3}. \tag{A.39}$$

Parameters

If ξ is known, the two parameters α and k are given by

$$k = (\lambda_1 - \xi)/\lambda_2 - 2, \qquad \alpha = (1+k)(\lambda_1 - \xi). \tag{A.40}$$

If ξ is unknown, the three parameters are given by

$$k = (1 - 3\tau_3)/(1 + \tau_3), \quad \alpha = (1 + k)(2 + k)\lambda_2, \quad \xi = \lambda_1 - (2 + k)\lambda_2. \tag{A.41}$$

A.6 Generalized extreme-value distribution

Definition

Parameters (3): ξ (location), α (scale), k (shape).

Range of x: $-\infty < x \le \xi + \alpha/k$ if $k > 0$; $-\infty < x < \infty$ if $k = 0$; $\xi + \alpha/k \le x < \infty$ if $k < 0$.

$$f(x) = \alpha^{-1} e^{-(1-k)y - e^{-y}}, \quad y = \begin{cases} -k^{-1} \log\{1 - k(x - \xi)/\alpha\}, & k \ne 0 \\ (x - \xi)/\alpha, & k = 0 \end{cases} \tag{A.42}$$

$$F(x) = e^{-e^{-y}} \tag{A.43}$$

$$x(F) = \begin{cases} \xi + \alpha\{1 - (-\log F)^k\}/k, & k \ne 0 \\ \xi - \alpha \log(-\log F), & k = 0 \end{cases} \tag{A.44}$$

Special cases: $k = 0$ is the Gumbel distribution; $k = 1$ is a reverse exponential distribution; that is, $1 - F(-x)$ is the cumulative distribution function of an exponential distribution.

Extreme-value distributions are often classified into three types with cumulative distribution functions as follows:

$$\text{type I}: \quad F(x) = \exp(e^{-x}), \quad -\infty < x < \infty, \tag{A.45}$$

$$\text{type II}: \quad F(x) = \exp(-x^{-\delta}), \quad 0 \le x < \infty, \tag{A.46}$$

$$\text{type III}: \quad F(x) = \exp(-|x|^{\delta}), \quad -\infty < x \le 0. \tag{A.47}$$

The generalized extreme-value distribution subsumes each of these types, types I, II, and III corresponding to $k = 0$, $k < 0$, and $k > 0$, respectively. The Weibull distribution defined by

$$F(x) = 1 - \exp[-\{(x - \zeta)/\beta\}^{\delta}], \quad \zeta \le x < \infty, \tag{A.48}$$

is a reverse generalized extreme-value distribution with parameters

$$k = 1/\delta, \quad \alpha = \beta/\delta, \quad \xi = \zeta - \beta. \tag{A.49}$$

L-moments

L-moments are defined for $k > -1$.

$$\lambda_1 = \xi + \alpha\{1 - \Gamma(1+k)\}/k \quad \text{(A.50)}$$

$$\lambda_2 = \alpha(1 - 2^{-k})\Gamma(1+k)/k \quad \text{(A.51)}$$

$$\tau_3 = 2(1 - 3^{-k})/(1 - 2^{-k}) - 3 \quad \text{(A.52)}$$

$$\tau_4 = \{5(1 - 4^{-k}) - 10(1 - 3^{-k}) + 6(1 - 2^{-k})\}/(1 - 2^{-k}) \quad \text{(A.53)}$$

Here $\Gamma(.)$ denotes the gamma function

$$\Gamma(x) = \int_0^\infty t^{x-1} e^{-t} dt\,. \quad \text{(A.54)}$$

Parameters

To estimate k, Eq. (A.52) must be solved for k. No explicit solution is possible, but the following approximation, given by Hosking et al. (1985b), has accuracy better than 9×10^{-4} for $-0.5 \le \tau_3 \le 0.5$:

$$k \approx 7.8590c + 2.9554c^2, \qquad c = \frac{2}{3+\tau_3} - \frac{\log 2}{\log 3}\,. \quad \text{(A.55)}$$

The other parameters are then given by

$$\alpha = \frac{\lambda_2 k}{(1 - 2^{-k})\Gamma(1+k)}\,, \qquad \xi = \lambda_1 - \alpha\{1 - \Gamma(1+k)\}/k\,. \quad \text{(A.56)}$$

A.7 Generalized logistic distribution

Definition

Parameters (3): ξ (location), α (scale), k (shape).

Range of x: $-\infty < x \le \xi + \alpha/k$ if $k > 0$; $-\infty < x < \infty$ if $k = 0$; $\xi + \alpha/k \le x < \infty$ if $k < 0$.

$$f(x) = \frac{\alpha^{-1} e^{-(1-k)y}}{(1+e^{-y})^2}\,, \qquad y = \begin{cases} -k^{-1}\log\{1 - k(x-\xi)/\alpha\}\,, & k \ne 0 \\ (x-\xi)/\alpha\,, & k = 0 \end{cases} \quad \text{(A.57)}$$

$$F(x) = 1/(1 + e^{-y}) \quad \text{(A.58)}$$

$$x(F) = \begin{cases} \xi + \alpha[1 - \{(1-F)/F\}^k]/k\,, & k \ne 0 \\ \xi - \alpha \log\{(1-F)/F\}\,, & k = 0 \end{cases} \quad \text{(A.59)}$$

Special cases: $k = 0$ is the logistic distribution.

This generalization of the logistic distribution differs from others that have been defined in the literature. It is a reparametrized version of the log-logistic distribution of Ahmad et al. (1988). The name is chosen to reflect the distribution's similarity to the generalized Pareto and generalized extreme-value distributions.

L-moments

L-moments are defined for $-1 < k < 1$.

$$\lambda_1 = \xi + \alpha(1/k - \pi/\sin k\pi) \tag{A.60}$$

$$\lambda_2 = \alpha k\pi/\sin k\pi \tag{A.61}$$

$$\tau_3 = -k \tag{A.62}$$

$$\tau_4 = (1 + 5k^2)/6 \tag{A.63}$$

Parameters

$$k = -\tau_3, \quad \alpha = \frac{\lambda_2 \sin k\pi}{k\pi}, \quad \xi = \lambda_1 - \alpha\left(\frac{1}{k} - \frac{\pi}{\sin k\pi}\right) \tag{A.64}$$

A.8 Lognormal distribution

Definition

Parameters (3): ξ (location), α (scale), k (shape).

Range of x: $-\infty < x \le \xi + \alpha/k$ if $k > 0$; $-\infty < x < \infty$ if $k = 0$; $\xi + \alpha/k \le x < \infty$ if $k < 0$.

$$f(x) = \frac{e^{ky - y^2/2}}{\alpha\sqrt{2\pi}}, \quad y = \begin{cases} -k^{-1}\log\{1 - k(x-\xi)/\alpha\}, & k \ne 0 \\ (x-\xi)/\alpha, & k = 0 \end{cases} \tag{A.65}$$

$$F(x) = \Phi(y) \tag{A.66}$$

$x(F)$ has no explicit analytical form

Here Φ is the cumulative distribution function of the standard Normal distribution, defined in Eq. (A.26).

Special case: $k = 0$ is the Normal distribution with parameters ξ and α.

The lognormal distribution is usually defined by

$$F(x) = \Phi[\{\log(x - \zeta) - \mu\}/\sigma], \qquad \zeta \le x < \infty. \tag{A.67}$$

Our reparametrization of the lognormal distribution in terms of ξ, α, and k is a small modification of the parametrization of Munro and Wixley (1970). It has several advantages over the usual parametrization using μ, σ, and ζ:

- within a single distribution it includes both lognormal distributions with positive skewness and a lower bound ($k < 0$), and lognormal distributions with negative skewness and an upper bound ($k > 0$);
- it includes the Normal distribution as a special case ($k = 0$) rather than as an unattainable limit;
- it exhibits the similarity in structure of the lognormal distribution to the generalized Pareto and generalized extreme-value distributions; and
- its parameters are more meaningful and more stable to estimate than are those of the standard parametrization of the distribution, particularly when the skewness is close to zero.

In our parametrization, the lognormal distribution is the distribution of a random variable X that is related to a random variable Z that has a standard Normal distribution, with mean 0 and variance 1, by

$$X = \begin{cases} \xi + \alpha(1 - e^{-kZ})/k\,, & k \ne 0, \\ \xi + \alpha Z\,, & k = 0. \end{cases} \tag{A.68}$$

The standard parametrization (A.67) may be obtained from our parametrization by setting

$$k = -\sigma, \qquad \alpha = \sigma e^{\mu}, \qquad \xi = \zeta + e^{\mu}, \tag{A.69}$$

and results for it may be derived from those below.

L-moments

L-moments are defined for all values of k.

$$\lambda_1 = \xi + \alpha(1 - e^{k^2/2})/k \tag{A.70}$$

$$\lambda_2 = \frac{\alpha}{k} e^{k^2/2}\{1 - 2\Phi(-k/\sqrt{2})\} \tag{A.71}$$

Table A.1. *Coefficients of the approximations (A.72)–(A.74).*

	$\tau_4^0 = 1.2260172 \times 10^{-1}$	
$A_0 = 4.8860251 \times 10^{-1}$	$C_0 = 1.8756590 \times 10^{-1}$	$E_0 = 2.0466534$
$A_1 = 4.4493076 \times 10^{-3}$	$C_1 = -2.5352147 \times 10^{-3}$	$E_1 = -3.6544371$
$A_2 = 8.8027039 \times 10^{-4}$	$C_2 = 2.6995102 \times 10^{-4}$	$E_2 = 1.8396733$
$A_3 = 1.1507084 \times 10^{-6}$	$C_3 = -1.8446680 \times 10^{-6}$	$E_3 = -0.20360244$
$B_1 = 6.4662924 \times 10^{-2}$	$D_1 = 8.2325617 \times 10^{-2}$	$F_1 = -2.0182173$
$B_2 = 3.3090406 \times 10^{-3}$	$D_2 = 4.2681448 \times 10^{-3}$	$F_2 = 1.2420401$
$B_3 = 7.4290680 \times 10^{-5}$	$D_3 = 1.1653690 \times 10^{-4}$	$F_3 = -0.21741801$

There are no simple expressions for the L-moment ratios τ_r, $r \geq 3$. They are functions of k alone and can be computed by numerical integration, as in Hosking (1996). Alternatively, rational-function approximations can be used. The following approximations for τ_3 and τ_4 have accuracy better than 2×10^{-7} and 5×10^{-7}, respectively, for $|k| \leq 4$, corresponding to $|\tau_3| \leq 0.99$ and $\tau_4 \leq 0.98$:

$$\tau_3 \approx -k \frac{A_0 + A_1 k^2 + A_2 k^4 + A_3 k^6}{1 + B_1 k^2 + B_2 k^4 + B_3 k^6}, \tag{A.72}$$

$$\tau_4 \approx \tau_4^0 + k^2 \frac{C_0 + C_1 k^2 + C_2 k^4 + C_3 k^6}{1 + D_1 k^2 + D_2 k^4 + D_3 k^6}. \tag{A.73}$$

The coefficients used in the approximations are given in Table A.1.

Parameters

The shape parameter k is a function of τ_3 alone. No explicit solution is possible, but the following approximation has relative accuracy better than 2.5×10^{-6} for $|\tau_3| \leq 0.94$, corresponding to $|k| \leq 3$:

$$k \approx -\tau_3 \frac{E_0 + E_1 \tau_3^2 + E_2 \tau_3^4 + E_3 \tau_3^6}{1 + F_1 \tau_3^2 + F_2 \tau_3^4 + F_3 \tau_3^6}. \tag{A.74}$$

The coefficients used in the approximation are given in Table A.1. The other parameters are then given by

$$\alpha = \frac{\lambda_2 k e^{-k^2/2}}{1 - 2\Phi(-k/\sqrt{2})}, \qquad \xi = \lambda_1 - \frac{\alpha}{k}(1 - e^{k^2/2}). \tag{A.75}$$

A.9 Pearson type III distribution

Definition

Parameters (3): μ (location), σ (scale), γ (shape).

If $\gamma \neq 0$, let $\alpha = 4/\gamma^2$, $\beta = \frac{1}{2}\sigma|\gamma|$, and $\xi = \mu - 2\sigma/\gamma$. If $\gamma > 0$, then the range of x is $\xi \leq x < \infty$ and

$$f(x) = \frac{(x-\xi)^{\alpha-1} e^{-(x-\xi)/\beta}}{\beta^\alpha \Gamma(\alpha)}, \tag{A.76}$$

$$F(x) = G\left(\alpha, \frac{x-\xi}{\beta}\right) / \Gamma(\alpha). \tag{A.77}$$

If $\gamma = 0$, then the distribution is Normal, the range of x is $-\infty < x < \infty$ and

$$f(x) = \phi\left(\frac{x-\mu}{\sigma}\right), \qquad F(x) = \Phi\left(\frac{x-\mu}{\sigma}\right). \tag{A.78}$$

If $\gamma < 0$, then the range of x is $-\infty < x \leq \xi$ and

$$f(x) = \frac{(\xi-x)^{\alpha-1} e^{-(\xi-x)/\beta}}{\beta^\alpha \Gamma(\alpha)}, \tag{A.79}$$

$$F(x) = 1 - G\left(\alpha, \frac{\xi-x}{\beta}\right) / \Gamma(\alpha). \tag{A.80}$$

In each case, $x(F)$ has no explicit analytical form. Here $\Gamma(.)$ is the gamma function, defined in Eq. (A.54), and

$$G(\alpha, x) = \int_0^x t^{\alpha-1} e^{-t} dt \tag{A.81}$$

is the incomplete gamma function. The functions $\phi(.)$ and $\Phi(.)$ are as defined in Eq. (A.26).

Special cases: $\gamma = 2$ is the exponential distribution; $\gamma = 0$ is the Normal distribution; $\gamma = -2$ is the reverse exponential distribution.

The Pearson type III distribution is usually regarded as consisting of just the case $\gamma > 0$ given above and is usually parametrized by α, β, and ξ. Our parametrization extends the distribution to include the usual Pearson type III distributions, with positive skewness and lower bound ξ, reverse Pearson type III distributions, with negative skewness and upper bound ξ, and the Normal distribution, which is included as a special case of the distribution rather than as the unattainable limit $\alpha \to \infty$. This enables the Pearson type III distribution to be used when the skewness of the observed data may be negative. The parameters μ, σ, and γ are the conventional moments of the distribution.

L-moments

Expressions for the distribution's *L*-moments in terms of its parameters are simpler when using the standard parameters; to present these results, we therefore use the standard parametrization, assuming $\gamma > 0$. The corresponding results for $\gamma < 0$ are obtained by changing the signs of λ_1, τ_3, and ξ wherever they occur in expressions (A.82)–(A.89).

L-moments are defined for all values of α, $0 < \alpha < \infty$.

$$\lambda_1 = \xi + \alpha\beta \tag{A.82}$$

$$\lambda_2 = \pi^{-1/2}\beta\Gamma(\alpha + \tfrac{1}{2})/\Gamma(\alpha) \tag{A.83}$$

$$\tau_3 = 6I_{1/3}(\alpha, 2\alpha) - 3 \tag{A.84}$$

Here $I_x(p, q)$ denotes the incomplete beta function ratio

$$I_x(p, q) = \frac{\Gamma(p+q)}{\Gamma(p)\Gamma(q)} \int_0^x t^{p-1}(1-t)^{q-1}dt\,. \tag{A.85}$$

There is no simple expression for τ_4. Rational-function approximations can be used to express τ_3 and τ_4 approximately as functions of α. The following approximations are accurate to 10^{-6}. If $\alpha \geq 1$,

$$\tau_3 \approx \alpha^{-1/2}\frac{A_0 + A_1\alpha^{-1} + A_2\alpha^{-2} + A_3\alpha^{-3}}{1 + B_1\alpha^{-1} + B_2\alpha^{-2}}, \tag{A.86}$$

$$\tau_4 \approx \frac{C_0 + C_1\alpha^{-1} + C_2\alpha^{-2} + C_3\alpha^{-3}}{1 + D_1\alpha^{-1} + D_2\alpha^{-2}}; \tag{A.87}$$

if $\alpha < 1$,

$$\tau_3 \approx \frac{1 + E_1\alpha + E_2\alpha^2 + E_3\alpha^3}{1 + F_1\alpha + F_2\alpha^2 + F_3\alpha^3}, \tag{A.88}$$

$$\tau_4 \approx \frac{1 + G_1\alpha + G_2\alpha^2 + G_3\alpha^3}{1 + H_1\alpha + H_2\alpha^2 + H_3\alpha^3}. \tag{A.89}$$

Coefficients of the approximations are given in Table A.2.

Table A.2. *Coefficients of the approximations (A.86)–(A.89).*

$A_0 = 3.2573501 \times 10^{-1}$	$C_0 = 1.2260172 \times 10^{-1}$
$A_1 = 1.6869150 \times 10^{-1}$	$C_1 = 5.3730130 \times 10^{-2}$
$A_2 = 7.8327243 \times 10^{-2}$	$C_2 = 4.3384378 \times 10^{-2}$
$A_3 = -2.9120539 \times 10^{-3}$	$C_3 = 1.1101277 \times 10^{-2}$
$B_1 = 4.6697102 \times 10^{-1}$	$D_1 = 1.8324466 \times 10^{-1}$
$B_2 = 2.4255406 \times 10^{-1}$	$D_2 = 2.0166036 \times 10^{-1}$
$E_1 = 2.3807576$	$G_1 = 2.1235833$
$E_2 = 1.5931792$	$G_2 = 4.1670213$
$E_3 = 1.1618371 \times 10^{-1}$	$G_3 = 3.1925299$
$F_1 = 5.1533299$	$H_1 = 9.0551443$
$F_2 = 7.1425260$	$H_2 = 2.6649995 \times 10^{1}$
$F_3 = 1.9745056$	$H_3 = 2.6193668 \times 10^{1}$

Parameters

To estimate α, Eq. (A.84) must be solved for α, replacing τ_3 by $|\tau_3|$ to enable a solution to be obtained when τ_3 is negative. The following approximation has relative accuracy better than 5×10^{-5} for all values of α. If $0 < |\tau_3| < \frac{1}{3}$, let $z = 3\pi\tau_3^2$ and use

$$\alpha \approx \frac{1 + 0.2906z}{z + 0.1882z^2 + 0.0442z^3}\,; \tag{A.90}$$

if $\frac{1}{3} \le |\tau_3| < 1$, let $z = 1 - |\tau_3|$ and use

$$\alpha \approx \frac{0.36067z - 0.59567z^2 + 0.25361z^3}{1 - 2.78861z + 2.56096z^2 - 0.77045z^3}\,. \tag{A.91}$$

Given α, the parameters of our preferred parametrization may be found from

$$\gamma = 2\alpha^{-1/2}\,\text{sign}(\tau_3), \quad \sigma = \lambda_2\pi^{1/2}\alpha^{1/2}\Gamma(\alpha)/\Gamma(\alpha + \tfrac{1}{2}), \quad \mu = \lambda_1\,. \tag{A.92}$$

A.10 Kappa distribution

Definition

Parameters (4): ξ (location), α (scale), k, h.

Range of x: upper bound is $\xi + \alpha/k$ if $k > 0$, ∞ if $k \le 0$; lower bound is $\xi + \alpha(1-h^{-k})/k$ if $h > 0$, $\xi + \alpha/k$ if $h \le 0$ and $k < 0$, and $-\infty$ if $h \le 0$ and $k \ge 0$.

$$f(x) = \alpha^{-1}\{1 - k(x-\xi)/\alpha\}^{1/k-1}\{F(x)\}^{1-h} \tag{A.93}$$

$$F(x) = [1 - h\{1 - k(x-\xi)/\alpha\}^{1/k}]^{1/h} \tag{A.94}$$

$$x(F) = \xi + \frac{\alpha}{k}\left\{1 - \left(\frac{1-F^h}{h}\right)^k\right\} \tag{A.95}$$

The cases $h = 0$ and $k = 0$ are included implicitly as the continuous limits of (A.93)–(A.95).

Special cases: $h = -1$ is the generalized logistic distribution; $h = 0$ is the generalized extreme-value distribution; $h = 1$ is the generalized Pareto distribution. The three-parameter kappa distribution of Mielke and Johnson (1973) is a special case of the kappa distribution defined here. Its cumulative distribution function

$$F(x) = (x/b)^\theta\{a + (x/b)^{a\theta}\}^{-1/a}, \qquad x \ge 0, \quad a, b, \theta > 0, \tag{A.96}$$

is obtained from Eq. (A.94) by taking $\xi = b$, $\alpha = b/(a\theta)$, $k = -1/(a\theta)$, and $h = -a$.

The kappa distribution is a four-parameter distribution that includes as special cases the generalized logistic, generalized extreme-value, and generalized Pareto distributions. The most useful range of parameter values is $h \ge -1$. Subject to this restriction, the L-moments of the distribution cover a large area of the (τ_3, τ_4) plane – see Figure A.1. For these reasons it is useful as a general distribution with which to compare the fit of two- and three-parameter distributions and for use in simulating artificial data in order to assess the accuracy of statistical methods.

L-moments

L-moments are defined if $h \ge 0$ and $k > -1$, or if $h < 0$ and $-1 < k < -1/h$.

$$\lambda_1 = \xi + \alpha(1 - g_1)/k \tag{A.97}$$

$$\lambda_2 = \alpha(g_1 - g_2)/k \tag{A.98}$$

$$\tau_3 = (-g_1 + 3g_2 - 2g_3)/(g_1 - g_2) \tag{A.99}$$

$$\tau_4 = (-g_1 + 6g_2 - 10g_3 + 5g_4)/(g_1 - g_2), \tag{A.100}$$

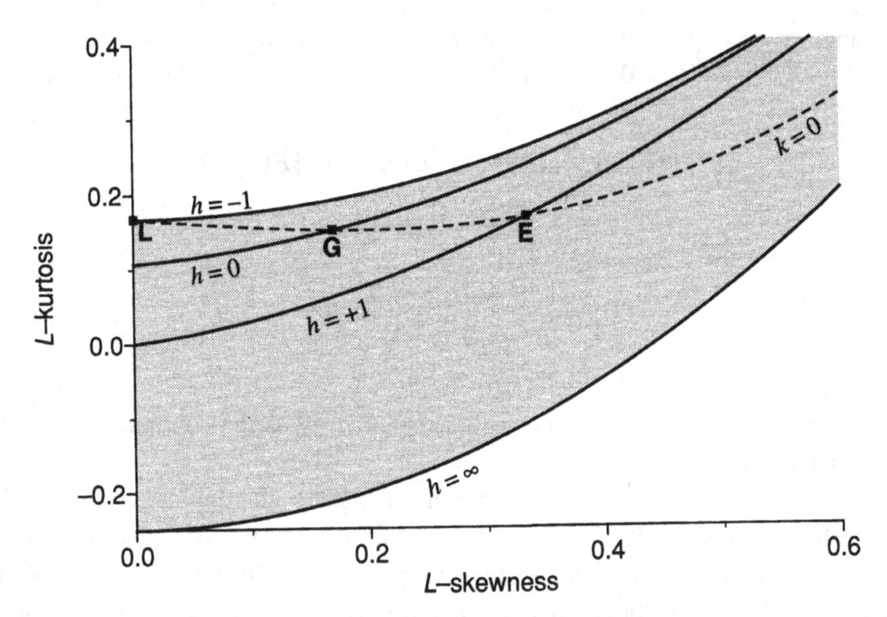

Fig. A.1. L-moment ratio diagram for the kappa distribution. The shaded area shows the L-skewness and L-kurtosis values attained by the kappa distribution with $h \geq -1$. Special cases include the logistic (L), Gumbel (G), exponential (E), generalized logistic ($h = -1$), generalized extreme-value ($h = 0$), and generalized Pareto ($h = +1$) distributions.

where

$$g_r = \begin{cases} \dfrac{r\Gamma(1+k)\Gamma(r/h)}{h^{1+k}\Gamma(1+k+r/h)}, & h > 0 \\[2ex] \dfrac{r\Gamma(1+k)\Gamma(-k-r/h)}{(-h)^{1+k}\Gamma(1-r/h)}, & h < 0. \end{cases} \tag{A.101}$$

Here $\Gamma(.)$ is the gamma function, defined in Eq. (A.54).

Parameters

There are no simple expressions for the parameters in terms of the L-moments. However, Eq. (A.99) and (A.100) give τ_3 and τ_4 in terms of k and h and can be solved for k and h given τ_3 and τ_4 by Newton-Raphson iteration. An algorithm is described by Hosking (1996).

A.11 Wakeby distribution

Definition

Parameters (5): ξ (location), α, β, γ, δ.

Range of x: $\xi \le x < \infty$ if $\delta \ge 0$ and $\gamma > 0$; $\xi \le x \le \xi + \alpha/\beta - \gamma/\delta$ if $\delta < 0$ or $\gamma = 0$.

$$f(x),\ F(x) \text{ not explicitly defined}$$

$$x(F) = \xi + \frac{\alpha}{\beta}\{1 - (1 - F)^{\beta}\} - \frac{\gamma}{\delta}\{1 - (1 - F)^{-\delta}\} \qquad \text{(A.102)}$$

Special cases: Both $\alpha = 0$ and $\gamma = 0$ give the generalized Pareto distribution.

The parametrization of the distribution is somewhat different from that used by other authors, for example, Landwehr et al. (1978, 1979b). Our parametrization explicitly exhibits the Wakeby distribution as a generalization of the generalized Pareto distribution and gives estimates of the α and γ parameters that are more stable under small perturbations of the data.

The distribution is invariant under the transformation $\alpha \leftrightarrow \beta$, $\gamma \leftrightarrow -\delta$, so without loss of generality we may assume that $\beta + \delta \ge 0$. Following Hosking (1986b) and Kotz, Johnson, and Read (1988, pp. 513–514), we assume that the following stronger conditions are satisfied:

(i) either $\beta + \delta > 0$ or $\beta = \gamma = \delta = 0$;
(ii) if $\alpha = 0$, then $\beta = 0$; and
(iii) if $\gamma = 0$, then $\delta = 0$.

For $x(F)$ to be a valid quantile function we must also impose the conditions

(iv) $\gamma \ge 0$; and
(v) $\alpha + \gamma \ge 0$.

The Wakeby distribution has been used by hydrologists to model streamflow data. The following properties of the distribution make it particularly suitable for applications in the environmental sciences:

- for suitable values of its parameters, the Wakeby distribution can mimic the shapes of many commonly used skew distributions (e.g., extreme-value, log-normal, Pearson type III);
- the Wakeby distribution has five parameters, more than most of the common distributions, and so can attain a wider range of distributional shapes than can the common distributions – this makes the Wakeby particularly useful for simulating artificial data for use in studying the robustness, under changes in distributional form, of methods of data analysis;
- when $\delta > 0$, the Wakeby distribution has a heavy upper tail and can therefore give rise to data sets containing occasional high outliers, a phenomenon often observed in environmental sciences;

- the Wakeby distribution has a finite lower bound, which is physically reasonable for many real-world observations; and
- the explicit form of the quantile function, Eq. (A.102), makes it easy to simulate random samples from the Wakeby distribution.

L-moments

L-moments are defined for $\delta < 1$.

$$\lambda_1 = \xi + \frac{\alpha}{(1+\beta)} + \frac{\gamma}{(1-\delta)} \tag{A.103}$$

$$\lambda_2 = \frac{\alpha}{(1+\beta)(2+\beta)} + \frac{\gamma}{(1-\delta)(2-\delta)} \tag{A.104}$$

$$\lambda_3 = \frac{\alpha(1-\beta)}{(1+\beta)(2+\beta)(3+\beta)} + \frac{\gamma(1+\delta)}{(1-\delta)(2-\delta)(3-\delta)} \tag{A.105}$$

$$\lambda_4 = \frac{\alpha(1-\beta)(2-\beta)}{(1+\beta)(2+\beta)(3+\beta)(4+\beta)} + \frac{\gamma(1+\delta)(2+\delta)}{(1-\delta)(2-\delta)(3-\delta)(4-\delta)} \tag{A.106}$$

There is no simple expression for τ_r.

Parameters

Landwehr et al. (1979b) gave an algorithm for parameter estimation using probability weighted moments. Hosking (1996) has implemented a variant of this algorithm that is expressed in terms of L-moments rather than probability weighted moments, does not impose Landwehr et al.'s arbitrary restriction on the range of the parameter β, and fits a generalized Pareto distribution if no Wakeby distribution is compatible with the given L-moments.

In terms of L-moments, the algorithm gives the following expressions for the parameters. If ξ is unknown, let

$$\begin{aligned} N_1 &= 3\lambda_2 - 25\lambda_3 + 32\lambda_4, & C_1 &= 7\lambda_2 - 85\lambda_3 + 203\lambda_4 - 125\lambda_5, \\ N_2 &= -3\lambda_2 + 5\lambda_3 + 8\lambda_4, & C_2 &= -7\lambda_2 + 25\lambda_3 + 7\lambda_4 - 25\lambda_5, \\ N_3 &= 3\lambda_2 + 5\lambda_3 + 2\lambda_4, & C_3 &= 7\lambda_2 + 5\lambda_3 - 7\lambda_4 - 5\lambda_5. \end{aligned} \tag{A.107}$$

Table A.3. *Polynomial approximations of τ_4 as a function of τ_3.*

	GPA	GEV	GLO	LN3	PE3	OLB
A_0	0.	0.10701	0.16667	0.12282	0.12240	−0.25
A_1	0.20196	0.11090	.	.	.	.
A_2	0.95924	0.84838	0.83333	0.77518	0.30115	1.25
A_3	−0.20096	−0.06669	.	.	.	.
A_4	0.04061	0.00567	.	0.12279	0.95812	.
A_5	.	−0.04208	.	.	.	.
A_6	.	0.03763	.	−0.13638	−0.57488	.
A_7	.	.	.	.	.	.
A_8	.	.	.	0.11368	0.19383	.

Note: The tabulated values are the coefficients of the polynomial approximation (A.115) for several distributions. Key to distributions: GPA – generalized Pareto, GEV – generalized extreme-value, GLO – generalized logistic, LN3 – lognormal, and PE3 – Pearson type III. OLB is the overall lower bound of τ_4 as a function of τ_3, given by Eq. (2.45).

Then β and $-\delta$ are the roots of the quadratic equation

$$(N_2C_3 - N_3C_2)z^2 + (N_1C_3 - N_3C_1)z + (N_1C_2 - N_2C_1) = 0, \qquad \text{(A.108)}$$

β being the larger of the two roots, and the other parameters are given by

$$\alpha = (1+\beta)(2+\beta)(3+\beta)\{(1+\delta)\lambda_2 - (3-\delta)\lambda_3\}/\{4(\beta+\delta)\}, \qquad \text{(A.109)}$$

$$\gamma = -(1-\delta)(2-\delta)(3-\delta)\{(1-\beta)\lambda_2 - (3+\beta)\lambda_3\}/\{4(\beta+\delta)\}, \qquad \text{(A.110)}$$

$$\xi = \lambda_1 - \alpha/(1+\beta) - \gamma/(1-\delta). \qquad \text{(A.111)}$$

If ξ is known, assume without loss of generality that $\xi = 0$ and let

$$\begin{aligned} N_1 &= 4\lambda_1 - 11\lambda_2 + 9\lambda_3, & C_1 &= 10\lambda_1 - 29\lambda_2 + 35\lambda_3 - 16\lambda_4, \\ N_2 &= -\lambda_2 + 3\lambda_3, & C_2 &= -\lambda_2 + 5\lambda_3 - 4\lambda_4, \\ N_3 &= \lambda_2 + \lambda_3, & C_3 &= \lambda_2 - \lambda_4. \end{aligned} \qquad \text{(A.112)}$$

Then β and $-\delta$ are the roots of the quadratic equation (A.108), β being the larger of the two roots, and the other parameters are given by

$$\alpha = (1+\beta)(2+\beta)\{\lambda_1 - (2-\delta)\lambda_2\}/(\beta+\delta), \qquad \text{(A.113)}$$

$$\gamma = -(1-\delta)(2-\delta)\{\lambda_1 - (2+\beta)\lambda_2\}/(\beta+\delta). \qquad \text{(A.114)}$$

A.12 Approximate L-skewness–L-kurtosis relationships

To construct an L-moment ratio diagram such as Figure 2.5, it is convenient to have simple explicit expressions for τ_4 in terms of τ_3 for commonly used probability distributions. Polynomial approximations of the form

$$\tau_4 = \sum_{k=0}^{8} A_k \tau_3^k \qquad (A.115)$$

have been obtained and the coefficients A_k are given in Table A.3. For given τ_3, the approximations yield values of τ_4 that are accurate to within 0.0005 over the range $-0.9 \le \tau_3 \le 0.9$, except that for the generalized extreme-value distribution, 0.0005 accuracy is obtained only when $-0.6 \le \tau_3 \le 0.9$. For the generalized Pareto distribution the exact relationship between τ_3 and τ_4 given by Eq. (A.39) can also be used.

A.13 L-moment ratio diagram

Figure A.2 is an L-moment ratio diagram covering the particularly useful range $0 \le \tau_3 \le 0.5$, $0 \le \tau_4 \le 0.4$. It is convenient for plotting sample at-site or regional average L-moment ratios for comparison with the population values of commonly used frequency distributions.

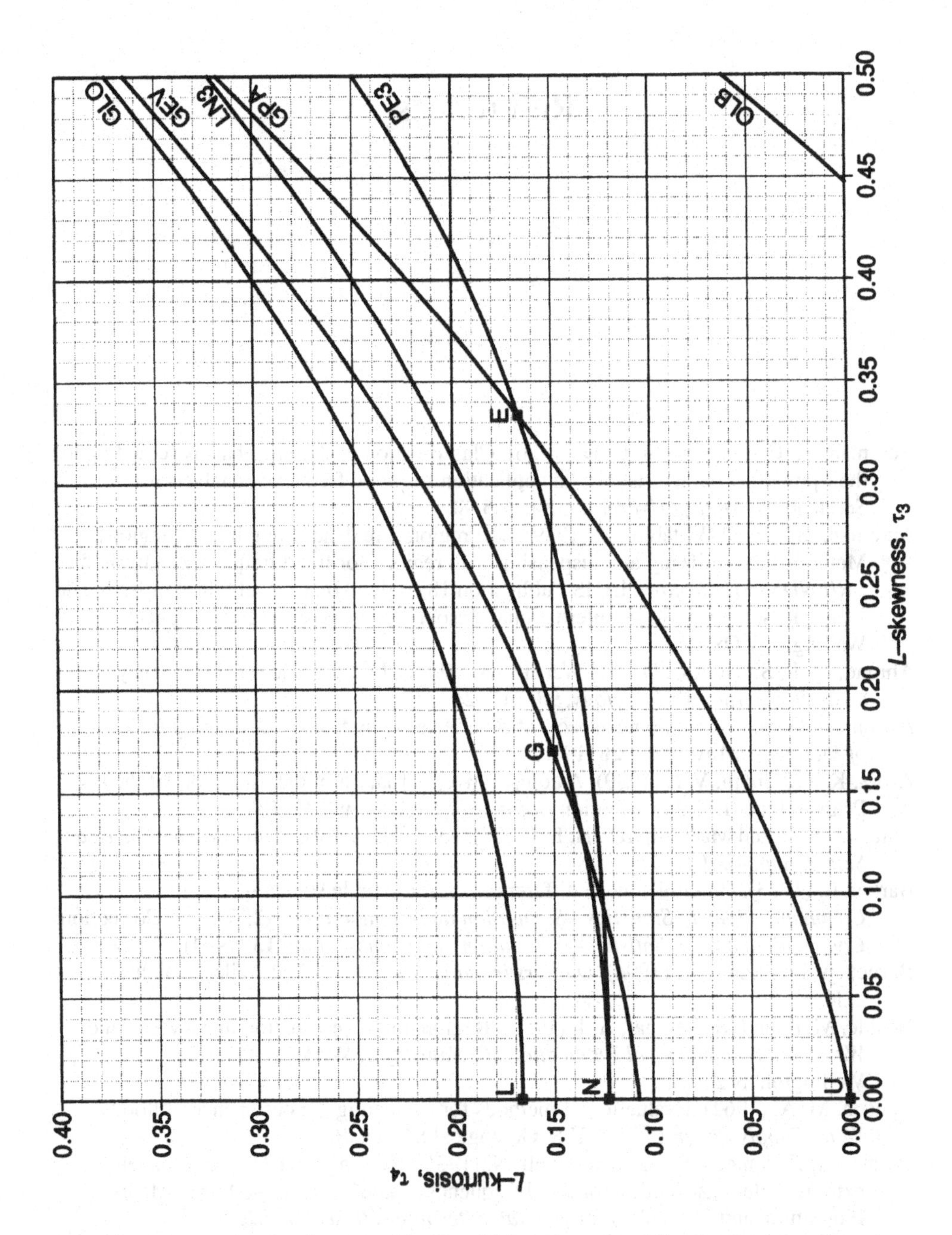

Fig. A.2. *L*-moment ratio diagram. Two- and three-parameter distributions are shown as points and lines, respectively. Key to distributions: E – exponential, G – Gumbel, L – logistic, N – Normal, U – uniform, GLO – generalized logistic, GEV – generalized extreme-value, GPA – generalized Pareto, LN3 – lognormal, and PE3 – Pearson type III. OLB is the overall lower bound of τ_4 as a function of τ_3, given by Eq. (2.45).

References

Acreman, M. C. and Sinclair, C. D. (1986). Classification of drainage basins according to their physical characteristics: An application for flood frequency analysis in Scotland. *Journal of Hydrology*, **84**, 365–80.

Acreman, M. C. and Wiltshire, S. (1989). The regions are dead: Long live the regions. Methods of identifying and dispensing with regions for flood frequency analysis. In *FRIENDS in Hydrology*, IAHS Publication 187, edited by L. Roald, K. Nordseth, and K. A. Hassel, pp. 175–88. International Association of Hydrological Sciences, Wallingford, Oxon.

Ahmad, M. I., Sinclair, C. D., and Werritty, A. (1988). Log-logistic flood frequency analysis. *Journal of Hydrology*, **98**, 215–24.

Alexander, G. N. (1954). Some aspects of time series in hydrology. *Journal of the Institute of Engineers (Australia)*, **26**, 196.

Arora, K. and Singh, V. P. (1989). A comparative evaluation of the estimators of the log Pearson type (LP) 3 distribution. *Journal of Hydrology*, **105**, 19–37.

Balanda, K. P. and MacGillivray, H. L. (1988). Kurtosis: A critical review. *The American Statistician*, **42**, 111–9.

Barker, L. (1983). On Gini's mean difference and the sample standard deviation. *Communications in Statistics—Simulation and Computation*, **12**, 503–5. (Correction: *Communications in Statistics—Simulation and Computation*, **13** (1984), 851–2.)

Barnett, V. and Lewis, T. (1994). *Outliers in statistical data*, 3rd ed. Wiley, Chichester, U.K.

Beable, M. E. and McKerchar, A. I. (1982). Regional flood estimation in New Zealand. *Water and Soil Technical Publication 20*, Ministry of Works and Development, Wellington, N.Z.

Benson, M. A. (1962). Evaluation of methods for evaluating the occurrence of floods. *Water Supply Paper 1550-A*, U.S. Geological Survey, Reston, Va.

Beran, M., Hosking, J. R. M., and Arnell, N. (1986). Comment on "Two-component extreme value distribution for flood frequency analysis" by Fabio Rossi, Mauro Fiorentino, and Pasquale Versace. *Water Resources Research*, **22**, 263–6.

Boes, D. C., Heo, J.-H., and Salas, J. D. (1989). Regional flood quantile estimation for a Weibull model. *Water Resources Research*, **25**, 979–90.

Boughton, W. C. (1980). A frequency distribution for annual floods. *Water Resources Research*, **16**, 347–54.

Buishand, T. A. (1989). Statistics of extremes in climatology. *Statistica Neerlandica*, **43**, 1–30.

Burn, D. H. (1988). Delineation of groups for regional flood frequency analysis. *Journal of Hydrology*, **104**, 345–61.
Burn, D. H. (1989). Cluster analysis as applied to regional flood frequency. *Journal of Water Resources Planning and Management*, **115**, 567–82.
Burn, D. H. (1990). Evaluation of regional flood frequency analysis with a region of influence approach. *Water Resources Research*, **26**, 2257–65.
Calinski, R. B. and Harabasz, J. (1974). A dendrite method for cluster analysis. *Communications in Statistics*, **3**, 1–27.
Caroni, C. and Prescott, P. (1992). Sequential application of Wilks's multivariate outlier test. *Applied Statistics*, **41**, 355 – 64.
Cavadias, G. S. (1990). The canonical correlation approach to regional flood estimation. In *Regionalization in Hydrology*, IAHS Publication no. 191, edited by M. A. Beran, M. Brilly, A. Becker and O. Bonacci, pp. 171–8. International Association of Hydrological Sciences, Wallingford, Oxon.
Chowdhury, J. U., Stedinger, J. R., and Lu, L.-H. (1991). Goodness-of-fit tests for regional generalized extreme value flood distributions. *Water Resources Research*, **27**, 1765–76.
Cohn, T. A. and Stedinger, J. R. (1987). Use of historical information in a maximum-likelihood framework. *Journal of Hydrology*, **96**, 215–23.
Cong, S., Li, Y., Vogel, J. L., and Schaake, J. C. (1993). Identification of the underlying distribution form of precipitation by using regional data. *Water Resources Research*, **29**, 1103–11.
Cong, S. and Xu Y. (1987). The effect of discharge measurement error in flood frequency analysis. *Journal of Hydrology*, **96**, 237–54.
Cox, D. R. and Hinkley, D. V. (1974). *Theoretical statistics*. Chapman and Hall, London.
Cunnane, C. (1978). Unbiased plotting positions—a review. *Journal of Hydrology*, **37**, 205–22.
Cunnane, C. (1988). Methods and merits of regional flood frequency analysis. *Journal of Hydrology*, **100**, 269–90.
Dalén, J. (1987). Algebraic bounds on standardized sample moments. *Statistics and Probability Letters*, **5**, 329–31.
Dales, M. Y. and Reed, D. W. (1989). Regional flood and storm hazard assessment. *Report 102*, Institute of Hydrology, Wallingford, U.K.
Dalrymple, T. (1960). Flood frequency analyses. *Water Supply Paper 1543-A*, U.S. Geological Survey, Reston, Va.
Damazio, J. M. and Kelman, J. (1986). Use of historical data in flood-frequency analysis. International Symposium on Flood Frequency and Risk Analysis, Louisiana State University, Baton Rouge, La., May 1986.
De Coursey, D. G. (1973). Objective regionalization of peak flow rates. In *Floods and Droughts*, Proceedings of the Second International Symposium in Hydrology, Fort Collins, Colorado, edited by E. L. Koelzer, V. A. Koelzer, and K. Mahmood, pp. 395–405. Water Resources Publications, Fort Collins, Colo.
Ding, J. and Yang, R. (1988). The determination of probability weighted moments with the incorporation of extraordinary values into sample data and their application to estimating parameters for the Pearson type three distribution. *Journal of Hydrology*, **101**, 63–81.
Duda, R. O. and Hart, P. E. (1973). *Pattern classification and scene analysis*. Wiley, New York.
Efron, B. (1982). *The jackknife, the bootstrap, and other resampling plans*, CBMS Monograph 38. SIAM, Philadelphia.

Ely, L. L., Enzel, Y., Baker, V. R. and Cayan, D. R. (1993). A 5000-year record of extreme floods and climate change in the southwestern United States. *Science*, **262**, 410–2.

Everitt, B. S. (1993). *Cluster analysis*, 3rd ed. Edward Arnold, London.

Farhan, Y. I. (1984). Regionalisation of surface water catchments in east bank of Jordan. In *Proceedings of the Symposium "Problems in Regional Hydrology," University of Freiburg*.

Fielding, A. (1977). Binary segmentation: the Automatic Interaction Detector and related techniques for exploring data structures. In *Analysis of Survey Data, Vol. 1*, edited by C. A. O'Muircheartaigh and C. Payne. Wiley, New York.

Fill, H. D. (1994). Improving flood quantile estimates using regional information. Ph. D. thesis, Cornell University, Ithaca, N.Y.

Fill, H. D. and Stedinger, J. R. (1995). Homogeneity tests based upon Gumbel distribution and a critical appraisal of Dalrymple's test. *Journal of Hydrology*, **166**, 81–105.

Fiorentino, M., Gabriele, S., Rossi, F., and Versace, P. (1987). Hierarchical approach for regional flood frequency analysis. In *Regional Flood Frequency Analysis*, edited by V. P. Singh, pp. 35 – 49. D. Reidel, Norwell, Mass.

Fovell, R. G. and Fovell, M.-Y. C. (1993). Climate zones of the conterminous United States defined using cluster analysis. *Journal of Climate*, **6**, 2103–35.

Gabriele, S. and Arnell, N. (1991). A hierarchical approach to regional flood frequency analysis. *Water Resources Research*, **27**, 1281–9.

Gerard, R. and Karpuk, E. W. (1979). Probability analysis of historical flood data. *Journal of the Hydraulics Division, American Society of Civil Engineers*, **105**, 1153–65.

Gingras, D. and Adamowski, K. (1993). Homogeneous region delineation based on annual flood generation mechanisms. *Hydrological Sciences Journal*, **38**, 103–21.

Gingras, D., Adamowski, K., and Pilon, P. J. (1994). Regional flood equations for the provinces of Ontario and Quebec. *Water Resources Bulletin*, **30**, 55–67.

Gini, C. (1912). Variabilità e mutabilità, contributo allo studio delle distribuzioni e delle relazione statistiche. *Studi Economico-Giuridici della Reale Università di Cagliari*, **3**, 3–159.

Gordon, A. D. (1981). *Classification: Methods for the exploratory analysis of multivariate data*. Chapman and Hall, London.

Greenwood, J. A., Landwehr, J. M., Matalas, N. C., and Wallis, J. R. (1979). Probability weighted moments: Definition and relation to parameters of several distributions expressable in inverse form. *Water Resources Research*, **15**, 1049–54.

Greis, N. P. and Wood, E. F. (1981). Regional flood frequency estimation and network design. *Water Resources Research*, **17**, 1167–77. (Correction: *Water Resources Research*, **19** (1983), 589–90.)

Gringorten, I. I. (1963). A plotting rule for extreme probability paper. *Journal of Geophysical Research*, **68**, 813–4.

Gumbel, E. J. (1958). *Statistics of extremes*. Columbia University Press, New York.

Guttman, N. B. (1993). The use of L-moments in the determination of regional precipitation climates. *Journal of Climate*, **6**, 2309–25.

Guttman, N. B., Hosking, J. R. M., and Wallis, J. R. (1993). Regional precipitation quantile values for the continental U.S. computed from *L*-moments. *Journal of Climate*, **6**, 2326–40.

Hartigan, J. A. and Wong, M. A. (1979). Algorithm AS 136: A *K*-means clustering algorithm. *Applied Statistics*, **28**, 100–8.

Hosking, J. R. M. (1986a). The theory of probability weighted moments. *Research Report RC12210*, IBM Research Division, Yorktown Heights, N.Y.

Hosking, J. R. M. (1986b). The Wakeby distribution. *Research Report RC12302*, IBM Research Division, Yorktown Heights, N.Y.

Hosking, J. R. M. (1989). Some theoretical results concerning *L*-moments. *Research Report RC14492*, IBM Research Division, Yorktown Heights, N.Y.

Hosking, J. R. M. (1990). *L*-moments: Analysis and estimation of distributions using linear combinations of order statistics. *Journal of the Royal Statistical Society, Series B*, **52**, 105–24.

Hosking, J. R. M. (1994). The four-parameter kappa distribution. *IBM Journal of Research and Development*, **38**, 251–8.

Hosking, J. R. M. (1995). The use of *L*-moments in the analysis of censored data. In *Recent Advances in Life-Testing and Reliability*, edited by N. Balakrishnan, pp. 545–64. CRC Press, Boca Raton, Fla.

Hosking, J. R. M. (1996). Fortran routines for use with the method of *L*-moments, Version 3. *Research Report RC* 20525, IBM Research Division, Yorktown Heights, N.Y.

Hosking, J. R. M. and Wallis, J. R. (1986a). Paleoflood hydrology and flood frequency analysis. *Water Resources Research*, **22**, 543–50.

Hosking, J. R. M. and Wallis, J. R. (1986b). The value of historical data in flood frequency analysis. *Water Resources Research*, **22**, 1606–12.

Hosking, J. R. M. and Wallis, J. R. (1987a). Parameter and quantile estimation for the generalized Pareto distribution. *Technometrics*, **29**, 339 – 49.

Hosking, J. R. M. and Wallis, J. R. (1987b). Correlation and dependence between annual maximum flood series. *Research Report RC12822*, IBM Research Division, Yorktown Heights, N.Y.

Hosking, J. R. M. and Wallis, J. R. (1988). The effect of intersite dependence on regional flood frequency analysis. *Water Resources Research*, **24**, 588–600.

Hosking, J. R. M. and Wallis, J. R. (1993). Some statistics useful in regional frequency analysis. *Water Resources Research*, **29**, 271–81. (Correction: *Water Resources Research*, **31** (1995), 251.)

Hosking, J. R. M. and Wallis, J. R. (1995). A comparison of unbiased and plotting-position estimators of *L*-moments. *Water Resources Research*, **31**, 2019–25.

Hosking, J. R. M., Wallis, J. R., and Wood, E. F. (1985a). An appraisal of the regional flood frequency procedure in the UK Flood Studies Report. *Hydrological Sciences Journal*, **30**, 85–109.

Hosking, J. R. M., Wallis, J. R., and Wood, E. F. (1985b). Estimation of the generalized extreme-value distribution by the method of probability-weighted moments. *Technometrics*, **27**, 251 – 61.

Houghton, J. C. (1978). Birth of a parent: The Wakeby distribution for modeling flood flows. *Water Resources Research*, **14**, 1105–9.

Hydrosphere (1993). *Hydrodata CD-ROMs, vol. 4.0: USGS Peak Values*. Hydrosphere Data Products, Boulder, Colo.

Jin, M. and Stedinger, J. R. (1989). Flood frequency analysis with regional and historical information. *Water Resources Research*, **25**, 925–36.

Kalkstein, L. S., Tan, G., and Skindlov, J. A. (1987). An evaluation of three clustering procedures for use in synoptic climatological classification. *Journal of Climate and Applied Meteorology*, **26**, 717–30.

Karl, T. R., Williams, C. N., Quinlan, F. T., and Boden, T. A. (1990). United States historical climatology network (HCN) serial temperature and precipitation data. *ORNL/CDIAC-30, NDP-019/R1*, Carbon Dioxide Information Analysis Center, Oak Ridge National Laboratory, Oak Ridge, Tenn.

Kendall, M. G. (1975). *Rank correlation methods*. Charles Griffin, London.
Kochel, R. C., Baker, V. R., and Patton, P. C. (1982). Paleohydrology of southwestern Texas. *Water Resources Research*, **18**, 1165–83.
Kotz, S., Johnson, N. L., and Read, C. B. (1988). *Encyclopedia of Statistical Sciences, vol. 9*. Wiley, New York.
Kuczera, G. (1992). Uncorrelated measurement error in flood frequency inference. *Water Resources Research*, **28**, 183–8.
Landwehr, J. M., Matalas, N. C., and Wallis, J. R. (1978). Some comparisons of flood statistics in real and log space. *Water Resources Research*, **14**, 902–20. (Correction: *Water Resources Research*, **15** (1979), 983–4.)
Landwehr, J. M., Matalas, N. C., and Wallis, J. R. (1979a). Probability-weighted moments compared with some traditional techniques in estimating Gumbel parameters and quantiles. *Water Resources Research*, **15**, 1055–64.
Landwehr, J. M., Matalas, N. C., and Wallis, J. R. (1979b). Estimation of parameters and quantiles of Wakeby distributions. *Water Resources Research*, **15**, 1361–79. (Correction: *Water Resources Research*, **15** (1979), 1672.)
Landwehr, J. M., Tasker, G. D., and Jarrett, R. D. (1987). Discussion of "Relative accuracy of log Pearson III procedures" by J. R. Wallis and E. F. Wood. *Journal of Hydraulic Engineering*, **113**, 1206–10.
Laursen, E. M. (1983). Comment on "Paleohydrology of southwestern Texas" by R. C. Kochel, V. R. Baker, and P. C. Patton. *Water Resources Research*, **19**, 1339.
Leese, M. N. (1973). Use of censored data in the estimation of Gumbel distribution parameters for annual maximum flood series. *Water Resources Research*, **9**, 1534–42.
Lettenmaier, D. P. and Potter, K. W. (1985). Testing flood frequency estimation methods using a regional flood generation model. *Water Resources Research*, **21**, 1903–14.
Lettenmaier, D. P., Wallis, J. R., and Wood, E. F. (1987). Effect of regional heterogeneity on flood frequency estimation. *Water Resources Research*, **23**, 313–23.
Lu, L.-H. (1991). Statistical methods for regional flood frequency investigations. Ph.D. thesis, Cornell University, Ithaca, N.Y.
Lu, L.-H. and Stedinger, J. R. (1992a). Sampling variance of normalized GEV/PWM quantile estimators and a regional homogeneity test. *Journal of Hydrology*, **138**, 223–45.
Lu, L.-H. and Stedinger, J. R. (1992b). Variance of two- and three-parameter GEV/PWM quantile estimators: Formulae, confidence intervals, and a comparison. *Journal of Hydrology*, **138**, 247–67.
McKerchar, A. I. and Pearson, C. P. (1990). Maps of flood statistics for regional flood frequency analysis in New Zealand. *Hydrological Sciences Journal*, **35**, 609–21.
McMahon, T. A. and Srikanthan, R. (1982). Log Pearson type 3 distribution effect of dependence, distribution parameters and sample size on peak annual flood estimates. *Journal of Hydrology*, **52**, 149–59.
Matalas, N. C., Slack, J. R., and Wallis, J. R. (1975). Regional skew in search of a parent. *Water Resources Research*, **11**, 815–26.
Mielke, P. W. and Johnson, E. S. (1973). Three-parameter kappa distribution maximum likelihood estimates and likelihood ratio tests. *Monthly Weather Review*, **101**, 701–7.
Moore, R. J. (1987). Combined regional flood frequency analysis and regression on catchment characteristics by maximum likelihood estimation. In *Regional Flood Frequency Analysis*, edited by V. P. Singh, pp. 119–31. Reidel, Dordrecht, Netherlands.
Mosley, M. P. (1981). Delimitation of New Zealand hydrological regions. *Journal of Hydrology*, **49**, 173–92.

Munro, A. H. and Wixley, R. A. J. (1970). Estimators based on order statistics of small samples from a three-parameter lognormal distribution. *Journal of the American Statistical Association*, **65**, 212–25.

Nathan, R. J. and McMahon, T. A. (1990). Identification of homogeneous regions for the purposes of regionalisation. *Journal of Hydrology*, **111**, 217–38.

Natural Environment Research Council (1975). *Flood Studies Report, vol. 1*. Natural Environment Research Council, London.

Neuman, C. P. and Schonbach, D. I. (1974). Discrete (Legendre) orthogonal polynomials: A survey. *International Journal for Numerical Methods in Engineering*, **8**, 743–70.

Pearson, C. P. (1991a). New Zealand regional flood frequency analysis using *L*-moments. *Journal of Hydrology (New Zealand)*, **30**, 53–64.

Pearson, C. P. (1991b). Regional flood frequency for small New Zealand basins, 2: Flood frequency groups. *Journal of Hydrology (New Zealand)*, **30**, 77–92.

Pearson, C. P. (1993). Application of L moments to maximum river flows. *The New Zealand Statistician*, **28**, 2–10.

Pilon, P. J. and Adamowski, K. (1992). The value of regional information to flood frequency analysis using the method of *L*-moments. *Canadian Journal of Civil Engineering*, **19**, 137 – 47.

Pilon, P. J., Adamowski, K., and Alila, Y. (1991). Regional analysis of annual maxima precipitation using *L*-moments. *Atmospheric Research*, **27**, 81–92.

Plackett, R. L. (1947). Limits of the ratio of mean range to standard deviation. *Biometrika*, **34**, 120–2.

Plantico, M. S., Karl, T. R., Kukla, G., and Gavin, J. (1990). Is recent climate change across the United States related to rising levels of anthropogenic greenhouse gases? *Journal of Geophysical Research*, **95**, 16617–37.

Potter, K. W. and Lettenmaier, D. P. (1990). A comparison of regional flood frequency estimation methods using a resampling method. *Water Resources Research*, **26**, 415–24.

Potter, K. W. and Walker, J. F. (1981). A model of discontinuous measurement error and its effect on the probability distribution of flood discharge measurements. *Water Resources Research*, **17**, 1505–9.

Potter, K. W. and Walker, J. F. (1985). An empirical study of flood measurement error. *Water Resources Research*, **21**, 403–6.

Reed, D. W. and Stewart, E. J. (1994). Inter-site and inter-duration dependence on rainfall extremes. In *Statistics for the Environment 2: Water Related Issues*, edited by V. Barnett and K. F. Turkman, pp. 125–43. Wiley, New York.

Richman, M. B. and Lamb, P. J. (1985). Climatic pattern analysis of three- and seven-day summer rainfall in the central United States: Some methodological considerations and a regionalization. *Journal of Climate and Applied Meteorology*, **24**, 1325–42.

Rosbjerg, D. and Madsen, H. (1995). Uncertainty measures of regional flood frequency estimators. *Journal of Hydrology*, **167**, 209–24.

Rossi, F., Fiorentino, M., and Versace, P. (1984). Two-component extreme value distribution for flood frequency analysis. *Water Resources Research*, **20**, 847–56.

Royston, P. (1992). Which measures of skewness and kurtosis are best? *Statistics in Medicine*, **11**, 333 – 43.

SAS (1988). *SAS/STAT user's guide*, Release 6.03. Cary, N.C.: SAS Institute.

Schaefer, M. G. (1990). Regional analyses of precipitation annual maxima in Washington State. *Water Resources Research*, **26**, 119–31.

Schroeder, E. E. and Massey, S. E. (1977). Techniques for estimating the magnitude and frequency of floods in Texas. *Report 77-110*, Water Resources Division, U.S. Geological Survey, Reston, Va.

Seaber, P. R., Kapinos, F. P., and Knapp, G. L. (1987). Hydrologic unit maps. *Water-Supply Paper 2294*, U.S. Geological Survey, Denver, Colo.

Simiu, E., Changery, M. J., and Filliben, J. J. (1979). Extreme wind speeds at 129 stations in the contiguous United States. *Building Science Series 118*, National Bureau of Standards, Washington, D.C.

Smith, J. A. (1989). Regional flood frequency analysis using extreme order statistics of the annual peak record. *Water Resources Research*, **25**, 311–7.

Smith, J. A. (1992). Representation of basin scale in flood peak distributions. *Water Resources Research*, **28**, 2993–9.

Stedinger, J. R. (1983). Estimating a regional flood frequency distribution. *Water Resources Research*, **19**, 503–10.

Stedinger, J. R. and Lu, L. H. (1995). Appraisal of regional and index flood quantile estimators. *Stochastic Hydrology and Hydraulics*, **9**, 49–75.

Stedinger, J. R. and Tasker, G. D. (1985). Regional hydrologic analysis 1. Ordinary, weighted and generalized least squares compared. *Water Resources Research*, **21**, 1421–32.

Stedinger, J. R. and Tasker, G. D. (1986). Regional hydrologic analysis 2. Model-error estimators, estimation of sigma and log-Pearson type 3 distributions. *Water Resources Research*, **22**, 1487–99.

Stedinger, J. R., Vogel, R. M., and Foufoula-Georgiou, E. (1992). Frequency analysis of extreme events. In *Handbook of Hydrology*, edited by D. R. Maidment, Chapter 18. McGraw-Hill, New York.

Stuart, A. and Ord, J. K. (1987). *Kendall's advanced theory of statistics, vol. 1*, 5th ed. Charles Griffin, London.

Tasker, G. D. and Stedinger, J. R. (1986). Regional skew with weighted LS regression. *Journal of Water Resources Planning and Management*, **112**, 225–37.

Tasker, G. D. and Thomas, W. O. (1978). Flood frequency analysis with pre-record information. *Journal of the Hydraulics Division, American Society of Civil Engineers*, **104**, 249–59.

Thompson, W. R. (1935). On a criterion for the rejection of observations and the distribution of the ratio of deviation to sample standard deviation. *Biometrika*, **32**, 214–9.

Tukey, J. W. (1960). The practical relationship between the common transformations of percentages of counts and of amounts. *Technical Report 36*, Statistical Techniques Research Group, Princeton Univ., Princeton, N.J.

U.S. Army Corps of Engineers (1992). HEC-FFA flood frequency analysis user's manual. *Report CPD-13*, Hydrologic Engineering Center, Davis, Calif.

U.S. Water Resources Council (1976). Guidelines for determining flood flow frequency. *Bulletin 17*, Hydrology Committee, Washington, D.C.

U.S. Water Resources Council (1977). Guidelines for determining flood flow frequency. *Bulletin 17A*, Hydrology Committee, Washington, D.C.

U.S. Water Resources Council (1981). Guidelines for determining flood flow frequency. *Bulletin 17B*, Hydrology Committee, Washington, D.C.

Vogel, R. M. and Fennessey, N. M. (1993). L-moment diagrams should replace product-moment diagrams. *Water Resources Research*, **29**, 1745–52.

Voss, R. F. (1995). *VossPlot*. Springer-Verlag, New York.

Wallis, J. R. (1981). Risk and uncertainties in the evaluation of flood events for the design of hydraulic structures. In *Piene e Siccità*, edited by E. Guggino, G. Rossi, and E. Todini, pp. 3–36. Fondazione Politecnica del Mediterraneo. Catania, Italy.

Wallis, J. R. (1982). Hydrologic problems associated with oilshale development. In *Environmental Systems and Management*, edited by S. Rinaldi, pp. 85–102. North-Holland, Amsterdam.

Wallis, J. R. (1993). Regional frequency studies using *L*-moments. In *Concise Encyclopedia of Environmental Systems*, edited by P. C. Young, pp. 468–76. Pergamon, Oxford.

Wallis, J. R., Lettenmaier, D. P., and Wood, E. F. (1991). A daily hydroclimatological data set for the continental United States. *Water Resources Research*, **27**, 1657 – 63.

Wallis, J. R., Matalas, N. C., and Slack, J. R. (1974). Just a moment! *Water Resources Research*, **10**, 211–9.

Wallis, J. R., Matalas, N. C., and Slack, J. R. (1977). Apparent regional skew. *Water Resources Research*, **13**, 159–82.

Wallis, J. R. and Wood, E. F. (1985). Relative accuracy of log Pearson III procedures. *Journal of Hydraulic Engineering*, **111**, 1043–56.

Wang, Q. J. (1990a). Estimation of the GEV distribution from censored samples by method of partial probability weighted moments. *Journal of Hydrology*, **120**, 103–14.

Wang, Q. J. (1990b). Unbiased estimation of probability weighted moments and partial probability weighted moments from systematic and historical flood information and their application to estimating the GEV distribution. *Journal of Hydrology*, **120**, 115–24.

White, E. L. (1975). Factor analysis of drainage basin properties: Classification of flood behavior in terms of basin geomorphology. *Water Resources Bulletin*, **11**, 676–87.

Wilkins, J. E. (1944). A note on skewness and kurtosis. *Annals of Mathematical Statistics*, **15**, 333–5.

Wilks, S. S. (1963). Multivariate statistical outliers. *Sankhyā*, **25**, 407–26.

Willeke, G. E., Hosking, J. R. M., Wallis, J. R., and Guttman, N. B. (1995). The National Drought Atlas (draft). *IWR Report 94-NDS-4*, U.S. Army Corps of Engineers, Fort Belvoir, Va.

Wiltshire, S. E. (1985). Grouping basins for regional flood frequency analysis. *Hydrological Sciences Journal*, **30**, 151–9.

Wiltshire, S. E. (1986a). Regional flood frequency analysis I: Homogeneity statistics. *Hydrological Sciences Journal*, **31**, 321–33.

Wiltshire, S. E. (1986b). Regional flood frequency analysis II: Multivariate classification of drainage basins in Britain. *Hydrological Sciences Journal*, **31**, 335 – 46.

Wiltshire, S. E. (1986c). Identification of homogeneous regions for flood frequency analysis. *Journal of Hydrology*, **84**, 287–302.

Zrinji, Z. and Burn, D. H. (1994). Flood frequency analysis for ungauged sites using a region of influence approach. *Journal of Hydrology*, **153**, 1–21.

Index of notation

Numbers in square brackets indicate the page on which the symbol is defined.

b_r	rth probability weighted moment of a data sample [26].
bias(.)	Bias of an estimator [16].
C_v	Coefficient of variation of a frequency distribution [17].
$\hat{C}_v$	Coefficient of variation of a data sample [18].
corr(., .)	Correlation between two random variables [15].
cov(., .)	Covariance of two random variables [15].
D_i	Discordancy measure for site i in a region [46].
E(.)	Expectation of a random variable [14].
F	Nonexceedance probability [2].
$F(.)$	Cumulative distribution function [14].
$f(.)$	Probability density function [14].
g	Skewness of a data sample [18].
H	Heterogeneity measure for regional data [63].
k	Kurtosis of a data sample [18]; k is also used as a parameter of several distributions.
ℓ_r	rth L-moment of a data sample [27].
$P_r^*(.)$	Shifted Legendre polynomial [20].
Pr[.]	Probability of an event [14].
Q_T	Quantile of return period T [2].
$Q_i(.)$	Quantile function at site i in a region [6].
$q(.)$	Regional growth curve [7].
RMSE(.)	Root mean square error of an estimator [16].
s	Standard deviation of a data sample [18].
^{T}	Transposition of a vector or matrix [46].
t	L-CV of a data sample [28].
t_r	rth L-moment ratio of a data sample [28].

var(.)	Variance of a random variable [15].
$X_{k:n}$	Random variable: the kth smallest element of a random sample of size n [21].
$x(.)$	Quantile function [14].
Z^{DIST}	Goodness-of-fit measure for regional data [81].
α_r	rth probability weighted moment of a frequency distribution [19].
β_r	rth probability weighted moment of a frequency distribution [19].
$\Gamma(.)$	Gamma function [196].
γ	Skewness of a frequency distribution [17]; Euler's constant [193]; γ is also used as a parameter of the Pearson type III and Wakeby distributions.
κ	Kurtosis of a frequency distribution [17].
λ_r	rth L-moment of a frequency distribution [20, 22].
μ	Mean of a frequency distribution [17].
σ	Standard deviation of a frequency distribution [17].
τ	L-CV of a frequency distribution [21].
τ_r	rth L-moment ratio of a frequency distribution [20].
$\Phi(.)$	Cumulative distribution function of the standard Normal distribution [193].
$\phi(.)$	Probability density function of the standard Normal distribution [193].

Author index

Subject index

Page numbers in **bold type** indicate formal definition of index terms.